T0155258

Counting Bounty

The Quest to Know the Worth of Earth

JEFFREY JOHNSON SMITH

County Bounty: The Quest to Know the Worth of Earth
Copyright ©2019/2020 Jeffery Johnson Smith. All Rights Reserved

Published by:
Trine Day LLC
PO Box 577
Walterville, OR 97489
1-800-556-2012
www.TrineDay.com
trineday@icloud.com

Library of Congress Control Number: 2020940981

Smith, Jeffery Johnson.
Counting Bounty – 1st ed.
p. cm.
Epub (ISBN-13) 978-1-63424-299-8
Kindle (ISBN-13) 978-1-63424-300-1
Print (ISBN-13) 978-1-63424-298-1
1. Income -- United States. 2. United States -- Economic policy. 3. United States
-- Economic conditions -- Statistics. 4. Land use -- United States. 5. Rent -- United States. 6. Real property -- Prices -- United States. I. Title

FIRST EDITION
10 9 8 7 6 5 4 3 2 1

Printed in the USA
Distribution to the Trade by:
Independent Publishers Group (IPG)
814 North Franklin Street
Chicago, Illinois 60610
312.337.0747
www.ipgbook.com

Publisher's Foreword

For want of a nail the shoe was lost.
For want of a shoe the horse was lost.
For want of a horse the rider was lost.
For want of a rider the message was lost.
For want of a message the battle was lost.
For want of a battle the kingdom was lost.
And all for the want of a horseshoe nail.

– Proverb, earliest use, 13th Century

A horse! a horse! my kingdom for a horse!
–King Richard III, William Shakespeare

What have we lost? Our way? Our heritage? Our sanity? The year 2020 started out okay, but in just thirty-one days, by the end of January, there was an impeachment trial, basketball phenom Kobe Bryant was dead, the UK formally withdrew from the European Union and a global health emergency emerged. The rest of the year doesn't look much better, the global health emergency has been re-classified as a global pandemic, which has limited sports, entertainment and social interactions ... and soon a U.S. presidential election.

TrineDay is very honored to publish *County Bounty: The Quest to Know the Worth of Earth* by Jeffery Johnson Smith. He brings to the table years of research, dilligence and thought that can be help us all understand what our future could be. One where the benefice of mankind underpins our humanity and helps us all care for our fellow human beings instead of a continual dog-eat-dog existence.

Let judgment run down as waters, and righteousness as a mighty stream.

Onward to the utmost of futures!
Peace,
R.A. "Kris" Millegan
Publisher
TrineDay
August 3, 2020

"In places where jobs disappear, society falls apart. The public sector and civic institutions are poorly equipped to do much about it. When a community truly disintegrates, knitting it back together becomes a herculean, perhaps impossible task. Virtue, trust, and cohesion – the stuff of civilization – are difficult to restore. If anything, it's striking how public corruption seems to often arrive hand-in-hand with economic hardship."

"Capital doesn't care about us."

"…increasing automation accompanied by social ruin. We must make the market serve humanity rather than have humanity continue to serve the market."

"The United States should provide an annual income of $12,000 for each American aged 18-64, with the amount indexed to increase with inflation."

– Andrew Yang, *The War on Normal People: The Truth About America's Disappearing Jobs and Why Universal Basic Income Is Our Future*

Glossary

Cadastral: (of a map or survey) showing the extent, value, and ownership of land, especially for taxation.

Geonomics: Geonomics or Georgism, named after Henry George (1839-1897), is a philosophy and economic ideology that holds that everyone owns what they create, but that everything found in nature, most importantly land, belongs equally to all of humanity.

Mulct: verb - extract money from (someone) by fine or taxation, noun - a fine or compulsory payment.

Noosphere: a postulated sphere or stage of evolutionary development dominated by consciousness, the mind, and interpersonal relationships (frequently with reference to the writings of Teilhard de Chardin).

Physiocrat: a member of an 18th-century group of French economists who believed that agriculture was the source of all wealth and that agricultural products should be highly priced. Advocating adherence to a supposed natural order of social institutions, they also stressed the necessity of free trade.

Propertarians: advocates of a political philosophy (Propertarianism, or proprietarianism) that reduces all questions of ethics to the right to own property.

Rent: payments for land, not for buildings, but also includes subsurface minerals and supra-surface electromagnetic frequencies, and not just solids like land but also liquids like water and gases like air; in sum, whatever is natural and has economic value. Parsing finer, rent is not only actual payments but also the imputed value of a location.

Rentier: a person living on income from property or investments.

CONTENTS

A Pandemic Preface

Most of us – that is, nearly all of us – consider a virus like covid-19 strong. At the other end of spectrum – the weak end – hides economic justice. But think again.

- Just how weak would be an absence of poverty? Some of the hardest hit demographics are poor people.

- How weak would be an absence of pollution? Contaminants both mutate genes and enfeeble immune systems; viruses love that.

- How weak would be stress-free, multi-generational families? The hardest hit group are the elderly warehoused alone in expensive facilities.

- How weak is a healthy diet, something your tax dollars do not support unlike a diet of packaged goods high in salt and sugar which your elected representatives do subsidize?

- How weak is a functional community in which neighbors not only know but identify with and support one another, as in amazing Roseto Pennsylvania (Chapter 41)? Healthier societies enjoy smaller, human-scale wealth gaps, and the worth of Earth can be used to close that gap, as does the oil dividend in Alaska (Chapter 40).

Justice can be powerful, but it's not on the agenda. Instead, "our" elected representatives are dishing out trillions to deep-pocket insiders. That's people who not only don't need it but who could afford to lose hundreds of thousands – what they might spend on a birthday party – and not even notice it. Most of the rest of us bide our tongues, getting a few bucks of hush money deposited into our accounts.

Even if the informal moratorium on mortgage payments may in some places become official, banks are not hurting, not after the latest favors, plus the huge profits garnered in recent years while ordinary people caught nary a drop trickling down.

Local governments and school districts – to the extent they rely on local site values – might have to devalue land and make do with less revenue. Later, as the ultra-endowed spend their swollen wealth, it'll trickle down and re-inflate land values. Meanwhile, however, many feel the squeeze.

If, as the Chinese say, crisis is danger plus opportunity, then let's take the Chicago mayor's advice and "never waste a good crisis." Those always first in line already have. We have some catching up to do.

Drastic is the new normal. Stay indoors. Stay apart. Don't go to work. Spend multi-trillions like there's no tomorrow – favors for all and everyone invited to the party. This stimulus is money our governments don't have but must borrow. Politicians are digging deeper the debt crater for future taxpayers.

In this unhinged new world order, a critical mass could actually take seriously fundamental reform. That'd be the sort of revenue policies that'd not only mitigate any crisis. They'd also work to everyone's benefit during stable times.

First, rather than replace lost revenue, have governments reduce their waste. Harvard puts government waste in the trillions, a huge percentage of their budgets (Ch 41), in the ballpark with the current round of handouts. Consider slashing these big-budget items:

- Usually, when localities cut costs, they cut low-paid jobs and leave the high-paid sinecures in place with nothing to do. This time, cut both.

- Legalize victim-less crime and cut police budgets. If you want to fund anything, fund what addicts themselves say they need, which costs much less.

- Along those lines, let kids determine curriculum and sell-off or re-purpose empty school buildings.

- Repeal red tape and the bureaucrats that go with it. Instead, reform limited liability and tort law. Let imposing risk – as by sending one's pollution downwind or downstream – become business for insurance companies. They know what standards to require of customers.

These four reforms would tremendously streamline the state. You can probably think of more good ideas. Now turn to funding the remaining budget.

During slowdowns, governments collect far fewer taxes on sales and income. Since those taxes are now close to useless and always counterproductive (Ch 41), this is a good time to get rid of those drags on any economy, especially a hobbled one, altogether.

Replace them with charging full annual market value for government-granted privileges, like corporate charters, utility franchises, land titles, patents and copyrights, etc. If we charge their market value and no more, then the privilege holder will always be able to pay. Right now, with values down, fees would be low. Later, they'd rise, as the economy would revive.

If you're pro-economic growth, read up on your economic history. What did San Francisco after its earthquake and fire a century ago and the Asian Tigers prior to their ascendancy have in common? They put into practice – to a degree – these tax shifts. Basically, they exempted improvements and levied locations. It worked then, it worked in New York City after World War I, and it would work now.

It'd work so well our steward – the state – would have a surplus (Ch 40). Seriously. Land values would be sky high, waste would be axed. Repeal welfare – corporate and ordinary – and its bureaucracy. Instead, pay citizens a dividend.

Further, any "plandemic" becomes farther fetched. Sure, power corrupts and most people can't believe the worst of the powerful (since we can't know for sure, I pass on passing judgment). Yet once we redirect revenue to everyone, we deprive the black-budget boys. They'd have to hold a bake sale to finance any shenanigan.

Turn from this sound yet unsolicited advice to the intellectual quest at hand. This slowdown, like all economic downturns, may lower the values of locations and resources. The final estimate we reached may no longer be current (Ch 38). But it will be on the nose again, and soon. Market economies are nothing if not cyclical. Don't hold your breath but don't worry, either. Our method was sound so our total will be found in the near future. And that Citizens Dividend? It'll be fatter than you can imagine in your wildest dreams (Ch 41).

Now, let us examine that geonomic method.

FOREWORD

A NATION'S NATURE – WORTH COUNTING

Get your facts first, then you can distort them as you please.
— Mark Twain, humorist, author, proto-geonomist

BE FOREWARNED

Got it all. Beauty. Wealth. Intelligence. That's our planet, Earth. A pinup, a model for calendars. And worth so many trillions, how *could* anyone count them all?

Why would anyone – outside of idle curiosity – count them? For several good reasons. The worth of Earth in America is a statistic you could put to excellent use:

1st: Earth's worth can indicate the economy's health. Our overall spending for parts of never-produced nature – locations primarily – is a surplus. The bigger the surplus, the better the economy is performing, like a field in bloom.

2nd: Across space, sub-sectors vary in value. You could compare regions, see the value of where pollution is high versus where it is low. You'd have another way to tell how much wrecking the ecosystem is costing you, and how big the payoff would be by producing goods and services cleanly, efficiently.

3rd: Over time, the total rises and falls. By tracking the flow, you can tell where the economy is headed. Indeed, while no big-name economist predicted the most recent recession, several who focus on the value of land (and resources) – "geonomists" – did (Ch 28). Savers and investors (just about all of us) would appreciate knowing what phase comes next.

A possible fourth reason is that, were paradigms to shift, society could redirect natural value into universally beneficial projects. For instance, Alaska uses a portion of oil revenue to pay residents a dividend; anyone have bills to pay? Aspen, Colorado uses a slice to make housing affordable; otherwise, even doctors can't afford to live in that ritzy ski resort (Ch 40).

TABOO TABULATION

Given the insights that a figure for the worth of Earth can yield, why is it not already known? Or easily found? Or the stuff of ballads? Would those failings have anything to do with that fourth reason, the possibility of redirecting those values away from present recipients? Wouldn't be the first time that power influenced research (Ch 2).

While some of us want the value of land, others want land to lack a price. Lovers of nature, the outdoors, wildlife, natural sciences and science in general consider Earth to be priceless. Those folks, understandably, don't want to see land as a commodity. Yet beware of what you wish for. Where land lacks a price tag, people take more than they need and use that recklessly, as in the Brazilian rainforest.

Others are fine with nature as commodity, even if indifferent to knowing its aggregate value. Our public agencies calculate the returns to those providing labor and those investing capital but not those owning land. Why? And why don't conventional economists demand to know how much we all spend for the nature we use? Or work such concepts into their theories since people paid for never-produced land respond differently from people paid for supplying their labor or capital?

BUT WHAT IT'S DONE FOR US LATELY?

Seeking to know land value does have an honorable lineage. Long ago, measuring the annual output – i.e., value – of one's territory paid off for all humanity. In the agrarian society of ancient Sumer in the Fertile Crescent, rulers wanted to assign the best lands to the best farmers, or to rotate the families so everyone got a shot at the most fertile fields. And later, to take a cut. So they *counted* their harvests.

For humanity, the payoff from keeping those first accounts was huge. People learned to turn their little pictures into letters and eventually became literate. Further, their accurate accounts of output enabled them to create and exchange tokens of value; *they invented money*. Money and literacy went on to shape much of civilization as we know it. Once again, knowing the worth of Earth could pay off big.

We're now off on an intellectual quest, going where mainstream economists fear to tread. We'll wade into the torrential flow of payments for locations like those in Manhattan and into the crawling creek of payments for land in Death Valley. As Carl Sagan said, *"Somewhere, something incredible is waiting to be known."*

CHAPTER 1

BOTH THE MEASURE AND THE MEASURING – OUTRAGEOUS

From there to here, and here to there, stunning things are everywhere.
– after Dr. Seuss

IS TRILLION THE NEW BILLION?

We humans are pushovers for size, for big numbers, too. Like … the world's richest person who, if spending $1,000 every hour, every day, awake or asleep, even if retired, would have to live over 10,000 years to spend it all.

Another big number you should be able to drop into casual conversation is the worth of Earth in America. That's the combined value of downtown locations, farmland, forests, oil and other minerals, airwaves, ecosystem services, etc. It's how much we spend for the nature we use. I'd guess several trillions in the US annually.

Trillions. After a while, they all sound alike – million, billion, trillion. But the difference is huge. A stack of a million one dollar bills would reach as high as a 35-story building. Shimmy up a trillion dollars and you're over a quarter of the way to the moon.

Let's pause for a caveat, if I may, for the sake of clarity and brevity. Herein, the term "rent" is the original usage, payments for land, not for buildings, from when landlords were not building lords but lords of the land. Now days, "land" is not limited to the surface but also includes subsurface minerals and supra-surface electromagnetic frequencies, and not just solids like land but also liquids like water and gases like air (when it becomes scarce); in sum, *whatever is natural and has economic value.*

Parsing finer, rent is not only actual payments but also the imputed value of a location. It'd get unwieldy to explain all that every time. "Rent" – with or without quotation marks – is a handy shortcut you'll be seeing a lot of.

A Magic Number?

In the late 1980s when Japan was booming, its real estate peaked. Then *if you could afford to buy the center of Tokyo – the grounds of the royal palace – you could afford to buy California four times.* Of course, that ballistic value was speculative, well above what any rational person would or could pay. Actual location was much less but still a lot, given Japan's population density.

However humongous land value is, it's easy to make. As Woody Allen said, 80% of success is just showing up. It's the presence of society that generates site value. *As population grows, this value keeps getting bigger, automatically.*

> *The demand for land is constantly growing as the population increases, and since its supply is finite, its price must increase over time.*
> – Robert Stammers, CFA, *Investopedia*, 2017 Dec 7

While Earth's worth is hefty, producing Earth is effortless. It was already here. Nobody you know made the land (unless you're really, really old or very well connected). None of us poured oil in the ground, pulled up Manhattan island, etc. Nobody has to be paid to create nature.

Hence economists refer to payments for land as a surplus. Such "rent" (technical meaning) is an unintended byproduct, like wild mushrooms. Receiving rent, as the Dire Straits sang, is "Money for nothin', chicks for free."

Recall that some governments tap this windfall. Alaska, Aspen CO, and Singapore use it to pay residents dividends. An extra thousand per year can come in handy (Ch 40).

Maybe not as useful as cash in hand, but also welcome are reliable forecasts, which knowing rising land value makes possible. As society prospers, people blow their land payments up into a bubble. By keeping an eye out for the peak, some clever guys foretold the mid-2000's bubble burst (see Ch 28).

If only the devastated millions of citizens had heeded the warning ... but they never heard it

Hindering the Counting

Despite the usefulness of this statistic, the worth of America is not published, unlike annual reports on the economy, or weekly reports on unemployment, or daily reports on the stock market. Why is that? Why did this line of inquiry never develop?

Political pressure is plausible. Spendy lots, pushing up housing cost, prompted Aspen to tap location value to fund housing assistance. When other towns in Colorado flirted with the reform, the remorselessly lobbied legislature outlawed it (Ch 39).

As the police say, the obvious explanation is usually the right one. Who wanted rent for themselves? No one in particular, everyone in general. Anyone owning land knows the less attention paid to its size the better.

Focusing on rent creates controversy. A PhD examining the flow of rents is as rude as would be a male glancing at a female's leg back in the Victorian Era when chaste matrons put skirts on the legs of their pianos. So our unearthing of Earth's worth (pardon the wordplay) doesn't look easy.

For some of us, our indifference is innocent enough. Rent is a number and we're *not especially fond of math*, even suffer "*math-o-phobia*." Nor of statistics. "Stories change people; statistics give them something to argue about," said Bernie Siegel, an American writer last century. The rest of us, though, appreciate the challenge. That those who could calculate it, don't, makes me even more curious. Or is ignorance bliss?

> *The worth of Earth and the human brain are a bad match. Sure, some humans don't want the total known. But also, brains aren't fully conscious.*

CHAPTER 2

MODERN "MAN" CAN'T SEE THE LAND

ALL THE DOO-DAH DAY

Denial ain't just a river in Egypt.
– Mark Twain, humorist, author, proto-geonomist

UNFAMILIAR CAN BE INVISIBLE

Somethings right in front of you, you don't easily detect.

If you're bouncing a basketball, *you might miss a gorilla.*

If you've *never seen a ship before, you might miss Columbus on the horizon.*

And if you no longer need to know, you forget. Probably you can't track a deer even though your ancestors could. Nor do you register "socially-generated surplus" even though your descendants will.

Because Eskimos have different words for different kinds of snow, they can instantly recognize each kind. Yet people in temperate climes don't notice the different types. Even if an Eskimo pointed out the differences, others could not easily detect them.

How many people speak Chinese? Zero. Just like zero speak European. Neither exists. People who speak Mandarin don't understand Cantonese and vice versa. We Westerners over-bundle.

Normally bundling's not fatal. Except when a picker does not differentiate between mushrooms and toadstools, the latter being poisonous. Our bundling of spending for land with spending for human-made stuff isn't fatal—just harmful.

Sometimes we don't know because we can't, or can't easily know. We have a blind spot.

Since we can't know something about everything, our brains are inherently conservative. Yet as Walter Lippmann, who coined "Cold War" and "stereotype" and won two Pulitzers, said a century ago, "Where all think alike, no one thinks very much."

Then we turn to a specialist. However, doing so gives officials the leeway to make claims that don't hold up and yet become widely accepted. Thus we think of homebuyers who're deeply in debt to banks as "owners," not as "owers." We joke about paying rent to banks and banks being the actual owners, so we know about the relationship. But the variant we automatically, unconsciously use is "owner."

Numbness Feels So Good

Meanwhile, the environment – natural and social – constantly throws up challenges to our problem-solving ability. What do we do about unaffordable housing? Any idea? Most of us learn only new facts that fit in old frames. If we have no category for a fact, we dismiss it. Like kinds of snow, or tracks of deer.

Yet closing our minds shuts out solutions. Long ago, blind-spotted *Europeans used a plow that caused them much backbreaking labor. However, Chinese farmers curved the blade differently so it would glide much easier – and increased their harvest.*

Not knowing a fact differs from not acknowledging one. If you're unfamiliar with an issue, you cannot easily judge it on its merits. Hence lone voices in the wilderness are often ignored.

However, doubt can be irrational, too. For instance, superiors more sensitive to cost than to science may doubt underlings. Recall the engineer who warned everyone the O-ring would fail, as it did, killing the astronauts aboard the Challenger. But there were other priorities.

Human brains are hardwired such that both knowing and not knowing please them. Somethings we just don't want to know; then we choose to be in denial. Via denial and blind spots we cheat ourselves of knowing parts of reality – consequences be damned.

Why Won't Watchdogs Bark?

Society's critics routinely reveal all sorts of facts and fancies about corporations, dynastic families, and government cover-ups. Yet our watchdogs don't bark at the oldest "unearned income" (unearned by an individual owner but earned by some entity)—rent for land. Their blind spot keeps them from sounding an alarm.

Most can not see land but only what's on it—unaffordable housing—missing:

- what rises in value is not housing—already built—but land. Further ...

11

• who does the raising are not sellers and landlords but buyers and tenants with more money to spend on locations.

Paying for the never-produced – versus actual products – is a phenomenon that's invisible to most of us. As Eskimos have several words for snow, English has many words for financial transactions – buy, sell, lease, rent, hire, pay, etc. But without a unique meaning for "rent," we don't see how it's special.

Most economists go AWOL when it comes to assessing rent. These academics don't request aggregates from public bureaucrats. And statisticians don't feel inspired to tabulate the total on their own. In the absence of measurements, the non-measured thing disappears – and with it a great indicator.

> *"Across markets, price-to-income ratios peaked in 226 of 382 metro areas between 2005 and 2009 (with 99 of those metros peaking in 2006)."*
> – Alexander Hermann, Harvard Joint Center for Housing Studies
> "Price-to-Income Rations Are Nearing Historic Highs,"
> September 13, 2018

What followed, of course, was recession.

Long after the dust settles, a Johnny-come-lately may note how rising rent culminates in collapse. This observation, coming a decade late, does most people no good, leaving savers and investors, not to mention governments, to fend for themselves. That most bureaucrats and academics ignore the danger, might that be an act of negligence worth a watchdog's bark?

Blocked by politics and blindness, new ideas often don't fare well with old ways of thinking. Yet obstacles can be surmounted, history shows. Then new knowledge can, as they say, set us free.

> Right under our collective nose is a way to divine econom-
> ic performance – we just need to do some accounting first.

CHAPTER 3

KNOWLEDGE WORTH KNOWING

Real knowledge is to know the extent of one's ignorance.
– Confucius, philosopher, proto-geonomist

POWER TO SEE DEEPER, BROADER

What enables thinkers like Albert Einstein, Henry Ford, and Ben Franklin? (All these standouts were proto-geonomists.) As important as their powerful intellect is their lack of inhibition. These scouts and others follow their curiosity, wherever it may lead, into the knowledge hinterlands. Einstein, for example, said he rode a beam of light around the universe to see what would happen – quintessential thinking outside the box.

Such unique difference-makers don't just expand humanity's body of knowledge. They also discover new ways for humans to view their world. In our current quest, instead of seeing reality as a place of scarcity and making a living as dog-eat-dog, finding a hefty total for the worth of Earth in America would paint a picture of bounty.

In general, by expanding our worldview we expand our capabilities. As scientist and reformer Roger Bacon (1214-1292) noted, "Knowledge is power." Today's quest for knowledge – how much is the value of all land – is a quest for power, too: to predict and to gauge social surplus.

KNOWING BOUNTY

Long ago in agrarian days, it was not just gadflies who sought fresh knowledge but monarchs, too. Their officials measured and recorded harvests. Not only ancient Sumer and Egypt (preface) but also a millennium ago in England. The Normans defeated the Anglo-Saxons, who were not so fresh from having just defeated the Danes. To know in detail what they'd won, the victorious administration commissioned statisticians of the day to tabulate the crown's tax base. The final cadastre they named the Domesday Book in 1086 ("domes" being the root of "domestic" as is "dome" or "roof" whence we derive "a roof over one's head"). It counted the output of the land and the assets on it..

No longer do governments tally land alone, leaving that value to F.I.R.E. (Finance, Insurance, & Real Estate). Some private specialists do tally. *The Economist* in "Land-shackled economies: The paradox of soil" (4 Apr 2015) showed that where economies perform bountifully, there land values soar (duh).

In the naturally abundant Pacific Northwest before the arrival of European Americans, the Native Indians found hunting and gathering so easy that they celebrated potlatches. They gave away their possessions, didn't hoard them, in order to gain stature. They could share because they saw plenty.

However much one might prefer to remain thinking inside the box today, an eye-popping sum for the value of land, natural resources, EM spectrum, etc, enlarges that box. Being aware of the torrent of rent lets citizens contemplate what's best to do with this socially-generated surplus – as did the tribe of Chief Seattle.

Perhaps better than most, the already powerful already get it. Since knowledge shapes the dominant paradigm, the powerful do what they can to shape that knowledge. They:

- *donate to universities* – citadels of research and knowledge,

- *own the major media* – disseminators of knowledge they choose, and

- *hire lobbyists* – peddlers of certain knowledge, not all.

Try thinking outside *their* box.

> *"The new strategic foundations behave as though they are entitled to make public policy, and they are not shy about it."*
> – "Beware Big Donors" by Stanley N. Katz;
> *The Chronicle Review*, March 25, 2012

KNOWING WHAT SPENDING SPURS

We expect to get what we pay for yet don't realize what else we all get. When we spend money to buy cars, computers, and vacations, we reward people for providing us with their labor and capital. But when we buy or lease land or an oil field, we don't. On one hand, paying producers motivates them to produce more goods and services. On the other hand, paying owners for never-produced land motivates them to invest in lobbying for more favors. Good to know how the real world works (Ch 12).

Owners and sellers don't raise prices and rents so much as competing buyers and tenants bid up what they pay for land (often misnamed housing). As folks spend more on land, they have less to spend on the output of producers. As land grows more valuable, they have less to spend on clothes, clam chowder, a hair cut, or other things that reward the efforts of your neighbors. Over time, *the imbalance becomes great enough that some of your neighbors declare bankruptcy.* Eventually, *a tipping point is reached and recession ensues.*

> *Housing IS the Business Cycle*
> – Edward E. Leamer, National Bureau of
> Economic Research, September 2007

Economists who focus on rent, watching its value versus the other two classical returns – wages and "interest" (or "profit," sort of) – let's call "geonomists." By tracking rent, they predict the booms and busts of the business cycle with impressive accuracy (Chapter 28). Which is nice to know, especially if you're managing your own investment portfolio.

KNOWLEDGE THAT KEEPS ON YIELDING

Accurately forecasting coming economic performance is huge, not just for the wallet but also the noosphere. Scientists do not claim to know something unless they can reliably get the results that their theories predict. Able to predict, geonomists could make economics into a science.

Once economics operates on a sound footing, we project that the sky's the limit for this branch of knowledge.

Today, knowledge expands faster than ever. It could expand faster still with the right popular mindset. Yet as more people grow curious, key academics and bureaucrats grow more resistant – at least at the beginning of this intellectual quest.

CHAPTER 4

SEEK ANSWERS VS REPEAT STOCK PHRASES

If what you say is not consequential, project that onto what is.

BURY OR PRAISE RENT?

In our modern era, it seems to many of us that land no longer matters. Who farms anymore, anyway? Rent being of no consequence is just common sense. However, as Bertrand Russell (another proto-geonomist) noted, common sense is often merely popular misunderstanding. This case is the opposite: unpopular understanding.

Even critics who know things like the salary gap between CEOs and workers, and reformers who know the losses from smog dismiss land and rent. They do so without any pressure from the establishment, despite rent being key to their issues. It's to capture the value of land and resources that humans do things like wage war, exploit and develop pristine nature, lobby for property tax caps, subsidize factory farms, etc.

Somewhat harder to deal with, if you're curious to know the worth of Earth in America, is expert misunderstanding. Even those who should know better assume the value of land and resources is insignificant as a sum and a factor. Conventional economists and mainstream statisticians trivialize and misconstrue natural rent.

IT'S ALL ACADEMIC

When teaching economics, professors compare rent to the portion of earnings of a talented person above the earnings of an average person. No, that's wages, however towering. Want an accurate analogy? Try urban drivers stuffing coins into a parking meter, paying for location.

Naysayers dismiss land as just another asset like capital. It does not matter to them that one, unlike the other:

- is fixed, can not be relocated,
- can not be reproduced like a new iPhone,
- required nobody's labor to come into existence, and

- is essential for life while the other is welcome for comfort.

Differences like these show why the two play widely disparate roles in economies. Academics explain economies are systems of rewards—just like training pets—yet blur.

CLASS BIAS?

Furthermore, knowing the success of their economy, a society might find a socially beneficial use for *this windfall now arriving in very few pockets.* But why fantasize? Our critics say, *despite evidence to the contrary,* that no single demographic gets the lion's share of such spending; it already goes to a majority.

Yet, even if rent were a minor phenomenon, how is that a criticism? Most professional economists spend their entire careers focused on such. Check out the table of contents of any peer-reviewed journal: consumption of chewing gum during holidays in rural counties, etc.

You can't blame the help, of course. Academics depend on donors and subsidies, bureaucrats squarely on government expenditure. And those who donate to universities and politicians are those who capture most rent.

Since rentiers do not provide any labor nor capital to produce land, the rent they get is something for nothing. Measuring this stream of reward draws attention to landlords growing rich in their sleep, as John Stuart Mill put it. Do his modern equivalents hinder the counting of natural surplus?

Economies are rewards systems – just like training pets. When we pay for some Earth, we do something different from when we pay for goods and services that somebody produced. Reward producers, they're likely to produce more. Reward absentee owners, what do they do? Hire an extra shift at a land factory? That's not happening but they may speculate even more, exploit more land and resources, and hire another lobbyist. Keeping the total unknown helps their cause (Ch 12).

Despite so much opposition, we curious gadflies may have some allies, wannabe reformers who'd find useful a figure for what we spend for the land we use.

"GREENS"

Imagine that we completely destroy the entire ecosystem—and survive. We'd have to do what it did, like deliver rain. Stuff like that gets costly. Robert Costanza tallied up what we'd have to pay the planet to perform the services it now performs for free (like create oxygen). He estimated the value of ecosystem services at $166 trillion annually, a value that fluctuates annually due to human stresses. That's for all Earth, not just America. Nor is that an exchange value; nature does not need to be paid to provide its services. Yet if environmentalists can receive Constanza's financial figure, why not ours, too?

There is a hurdle. Loving the land, environmentalists cringe when seeing others treat it as an object of speculation. Our seeking a figure for the rental value of nature seems suspiciously speculative. Yet emotion aside, a figure for how much we all spend now for land somewhat degraded must make some wonder how much land in better health would then be worth.

It's not that humans *want* to foul their nest. A fouled nest is just a by-product of humans grasping for some of the worth of Earth. Because payments for land total so much, they attract lots of rational investors. A few of those investors and owners make money by putting nature to good use – selective logging, organic gardening, etc. Most make money by putting nature to bad use – clear-cutting, aquifer-draining, etc.

If you think you can live with despoliation and feel that only money matters, here's a fact to face: the financial cost from pollution-related death, sickness, and welfare is some $4.6 trillion in annual losses – or about 6.2% of the global economy.

Equally as damaging as misuse of land has been nonuse. Specifically, when owners underuse urban land, they displace others who then overuse suburban land. Farmland lost to asphalt and concrete might be better left as a source of food to feed urban dwellers. To infill cities, making them more compact, many environmental groups have endorsed the notion of public recovery of land values (Ch 39). In general, however, just like economists those "greens" might be curious to know how much would flow into public treasuries.

HOUSING REFORMERS SEE SITE RENT

The notion of infilling cities both attracts and repels urban advocates, wanting more housing but wanting it affordable; newcomers downtown can't help but bid up the rent. Like, the artistic element makes a warehouse district hip and cool. The caché attracts new residents with deeper pockets. Artists can't afford higher rents so they resettle elsewhere. It's like bees swarming from an old nest to a new, buildable site. (Was the trigger also speculator bees bidding up site values in the old hives?)

Those who move out leave behind old friends, neighbors, nearby family, and coworkers, stretching – even severing – those ties. Those who stay put then deal with not only with higher housing expenses and property taxes but new neighbors and an altered character of their neighborhood, too. Yet citizens and cities both are both better off when communities are stable.

It's the value of land that pushes up the price of housing. Urbanists might like to know the size of that trigger.

The state of the housing market starts to make more sense when you factor in land values.
— by Mat Spasic for *Markets & Money*

Reformers concerned about unaffordable housing might lend a hand in ferreting out a stat for rent, and they are many and well placed:

- A pair of academics in the Federal Reserve Bank of Boston's **New England Public Policy Center** finger land rent: *"inelastic land supply in some attractive locations, combined with the growing number of high-income families nationally, can partially explain the growing differences in house prices and incomes among cities."*

- **Harvard** reports that "Over 38 million American households can't afford their housing, an increase of 146% in the past 16 years."

- **Stanford** researchers found affordable housing raises local land value. By how much? Maybe they'll help us find out.

- **UC Berkeley** researchers found that more than half of low-income households in the Bay Area, including those in higher-income neighborhoods, are at risk of, or already experiencing, gentrification.

- A recent **University of Southern California** survey said the high cost of living, especially housing, which is location, was making it difficult for businesses to retain employees

- The **Urban Land Institute** is "Facing the Challenges of Affordable Senior Housing."

- *The Dirt* by the **American Society of Landscape Architects** notes low-paying but essential jobs downtown *don't* pay workers enough to live downtown.

- **Demographia** International Housing Affordability Survey: 2016

- **CityLab** reports the nation's largest generation, Millennials, have been slower to buy houses due to affordability and location; the Boomers' suburbs lack the appeal of urban housing.

- **BuildZoom** notes the high cost of housing in expensive coastal metros is driven not by construction costs but by land. Because developers must acquire spendy land, building on it and selling what gets built is not necessarily lucrative.

- **The Guardian**: Not the Housing Rights Committee but a techie making six figures who no longer can afford Silicon Valley who said venture capitalists should quit investing in silly apps and instead solve social issues.

- **Bloomberg** opined that tech corps should argue less about visas and more about affordable housing for the employees.

- The **Washington Post Wonk Blog** reported that the National Low Income Housing Coalition sees the gap between wages and rent growing. Later they reported that Hawaiians are fleeing paradise because rent is too high.

- **Mercury News** reports techies flee to places like Portland and Austin, making those places unaffordable. Or stay behind; in LA, more Millennials live with their parents than all-ages live in Chicago (nearly 4 million).

- **Politico** reports the YIMBY Party, a coalition of renters, blocked a San Francisco ballot initiative to move zoning decisions beyond local control. Frustrated by high rents, they call for higher density housing and the federal government to get involved. Perhaps the US would tally land values for us all.

- **Slate** in their section Metropolis: Cities of Today and Tomorrow, ran "San Francisco's Civil War." Across the bay in Oakland, some have turned to living in shipping containers with their tiny footprint on land.

- **Wired** reports tech workers do well financially and bid up housing prices, which takes a toll on everyone else, especially blue-collar and service workers, who are increasingly priced out.

The McKinsey Global Institute found that, by spending so much of their incomes on the rent or mortgage, households don't spend enough on consumer goods. Besides not stimulating others to produce, add in shortage of affordable housing, inflating its price. Mal-housing costs the economy between $143 billion and $233 billion annually, not taking into account second-order costs to health, education and the environment.

That's nearly a score of institutes, researchers, and press groups who might join the chorus calling for official number-crunchers to tabulate a total for land and other assets non-produced, and that's accurate.

If some agency would do that and release realistic figures for rent, the stat would not only supply environmentalists and urbanists with ammo, but when the running total peaks then everybody would have an indicator of the approaching downturn. Rational actors could take protective measures, re-arrange their portfolios. Meanwhile, you would not believe the estimates experts pass off now as somehow connected to reality. What gets into those guys, anyway?

CHAPTER 5

OFFICIAL STAMP OF INSIGNIFICANCE

Statistics are like a bikini. What they reveal is suggestive, but what they conceal is vital.

– Aaron Levenstein

BAIT & SWITCH?

Officials do post a number for the worth of Earth in America that's so microscopic, you're left wondering, What country are they referring to? The one with the biggest economy on the planet? Really? That's the best the biggest can do?

Your publicly funded Bureau of Economic Analysis (BEA) statisticians do allude to the value of land, by supplying a figure for rent for buildings. While in colloquial parlance rent refers to payments for temporary use of a building, in economics usage it means payment for impermanent use of land. You might expect specialists to use the technical definition, not the colloquial one. Everything else at official sites is in jargon.

Advertising land rent yet using building rent is the old bait-and-switch. Rather than inform the curious, the minuscule figure trivializes the notion of rent. If our spending for land and resources is so tiny, why bother paying it any attention? Non-critical laypeople just go with the flow – if it's official, assuredly it's accurate, right?

A negligible number deprives society of information it needs. OTOH, a realistic worldview lets people plan. Seeing rent peak, people can confront a gathering storm. Seeing a trough, people can buy low and save. Yet public servants don't deliver.

You never want to think poorly of others, but what's going on here? All this is so brazen, it's hard to swallow. It's like public statisticians knew the answer they wanted before they did the research.

FIG LEAF OR FIGURE?

Their official figure has other problems, too. *The bureaucrats don't give the cumulative total but a net.* Crazier still, *they base it not on accounting but on responses from owners* who pay less tax by reporting a lower figure. Furthermore, it's only for persons. Yet however much is rent paid to nonpersons – to corporations, foundations, governments, etc – it can't be much, since officially total corporate profit was only about 1/8th of total income.

> *He who defines the terms, wins the argument.*
> – Confucius, proto-geonomist

Anyway, playing with the cards we're dealt, BEA rent in 2017 Q3 was $0.7 trillion. Total income was $16.7 trillion, so their rent was about 4%. As paltry as 4% is, some years they tell the public that rent is even paltrier. In 2000, it was more like 2%. Further, they didn't put their stat on the bottom line but beneath it in an inconspicuous footnote.

That's their story and they're sticking to it – or not. At another BEA table for the same year it's smaller. The table is the National Income Product Account. Over the years various economists – Simon Kuznetz 1934, William Nordhaus and James Tobin 1972 (National Academy of Sciences 2005) – have said NIPA is incomplete and misleading yet it is the most important measure of economic activity for a nation.

Official total returns from the other two factors in production – wages to labor and "interest" to capital – are huge. How could rent for land be exponentially smaller? Not be one third of total income, but some years only one *thirty-third*? That defies common sense.

For the statistics that our number-crunchers *do* want you to take seriously, they pull out all the stops.

- Weekly reports on GDP.
- Reams of articles in journals on inflation.
- Conferences and prizes on unemployment.

Yet none of their pet stats predict the business cycle. Or reveal the economy's bounty. Rather they hide the pea under the shell.

Academics take their cue and leave unanalyzed an "insignificant" figure. (Economists are not the boldest people in the world. None would be the first to say something like a physicist stating space-time is curved.)

PALTRY PLEASES

What gives our bureaucrats the chutzpah to publish a trifling stat? Say they did provide a reliable figure for society's spending on the nature it uses. If it's hefty, a portion of the public would take an interest, contrary to the interests of present beneficiaries.

Probably land is so remunerative, it motivates rentiers to let their wishes be known. Down the line, underlings discretely discourage number-crunchers from whipping out their calculators. It is kowtowing but to those powerful enough to call the shots.

Money does influence bureaucracies.

- The FDA approves a risky pill for a pharmaceutical.
- The NRC issues a nuclear plant permit for a utility.
- Clinical doctors "proved" smoking was healthy.

Not incompetence, just misleading.

The actual statisticians who purvey unimportance of rents, how do they feel about it? If they are also professors, would they teach their students to distort? If a student had done that unbidden, what grade would the professor have given for such sloppy, misleading work?

As Shakespeare said, *"All's fair in love and war,"* and Timothy said, *"the love of money is the root of all evil."* Evil might be a stretch, but certainly not "all is fair." It's disturbing. Either our public servants are flawed, or my power of reason is flawed – or both. Whether it's incompetence or intentional fudging, neither justifies our public servants' huge public budget.

One thing for sure: it's a win for those who find public ignorance to be their private bliss.

CHAPTER 6

OUR IGNORANCE IS WHOSE BLISS?

A philosopher: Which is worse: ignorance or apathy?
A student: I don't know and I don't care.
The gentry: Whichever. Just don't change.

WHY BLIND?

We Yanks were raised on TV, males especially on the cowboy identity and the use of guns in defense of one's spread. *Our understanding of commons gave way long ago*, as did our awareness of commonwealth. Now it's the individual who's entitled, not community, so move along little dogie, don't poke your nose into rent around here.

Many don't know and don't want to know and prefer others don't know the worth of Earth in America. But the stat would be interesting and useful. So why the attitude?

One of our blinders is ownership. While the American Dream began as having a house to live in forever, it no longer means settling down for long. Now "The Dream" is less about owning a home, but more about cashing in big-time.

That's so even though the profit comes from the location, not from the buildings. Houses depreciate. *What appreciates is never produced land*, due to the growing presence of society. That cash is something for nothing, a fact that counting it would highlight.

While many people dimmed their lights, on its own, land disappeared from view.

- In the Industrial Era, most families moved to the city from the country. About 99% of the population quit desiring land in order to farm (Ch 6). Since hardly anyone farms now, many modern metropolitans have forgotten how much land matters.

- In America, a majority own land and the home upon it (more precisely, owe a mortgage). Our homes block the view of half of real estate – the land.

- Most people are not landlords or lenders – only *2/3 of 1%* lease out buildings. The *hoi polloi* are familiar with *paying* rent and mortgages but have no experience with *receiving* rent or mortgage interest payments.

- Most don't own the most expensive land – commercial land, especially downtowns – where locational values are gargantuan and rent really flows.

If anyone prefers keeping people in the dark about rent, they caught a demographic break.

CURIOSITY CHILLS THE CHAT

Popular years ago were paintings that were two in one. One was the familiar two-dimensional image, the other was a hidden three-dimensional image that emerged only after looking at the flat one in a certain way. When the 3-D one emerged, viewers squealed. However, not everyone could let their vision adjust to register the 3-D.

"Oh look, there's a dinosaur!" Another would say, *"Where? I don't see it."* If you declared, *"There, see the biplane?"* the frustrated doubter could say, *"You're trying to put me on."* If you kept at it: *"Wow, can't you see the angel?"* the mad blind-spotted man might say, *"Liar, there's nothing there!"*

Our payments to landowners, being money for nothing, may be why economists discretely avert their eyes; what's unearned becomes invisible. From their POV, why should others make a big deal about something that's not important? If others do acknowledge rent's role, and if that irritates mainstream economists, then those annoyed could damage the reputation of colleagues who can detect rent.

Avoiding hard questions reinforces normalcy bias – How things are is how they *should* be. After decades of not researching property, absentee ownership, or rents congealing into fortunes, etc, now economists only write about non-prying topics like household debt, trade, derivatives, etc. Such safe phenomena don't explain the inner workings of economies.

By omitting the factor of land from their analyses and conclusions, economists tacitly assert that all spendings motivate equally. However, while our spending for things that other humans provide grows the pie, our expenditure for land or resources is how some hog the pie. Not distinguishing these two spending types, economists miss how economies work, why they sometimes don't, and what to do it about. And by not seeing rent, of course most never measure it.

Economists Don't Have It Easy

Much more than other social studies, what economists examine is inherently political. To do economics thoroughly, practitioners must explain or ignore:

- some people are doing the work, but others care apturing the wealth;
- some people are paying taxes, but others are getting subsidies;
- some are paying tuition, others are endowing universities.

Most economists don't pry into the potentially controversial, like measuring rent; it could put their job in jeopardy. Most play it safe – sort of like some medical researchers skirting alternatives to pills and scalpels. They know their place.

Putting caution before curiosity, better let those sleeping dogs lie.

What We Don't Know Does Hurt Us

The powerful prefer widespread ignorance:

- Some politicians censor research into stem cells.
- The first thing some revolutionary governments do is shut down the universities.
- Like specialists everywhere, academics act out the priesthood syndrome. They, and nobody else, can define the breadth and depth of economics.

The body of knowledge is held back.

When rent is overlooked, also is: who gets it, how they get it, whether they deserve it, what they do with it, who gets left out, and who creates it – potentially, lots of inconvenient truths for everyone from homeowners and absentee owners to lenders and speculators.

As goes academia, so goes bureaucracy. Call the government's info line. *Nada.* Why bother, since they receive no requests from academics? Hence the statistical blackout. Never counting society's spending for locations keeps that stream of socially-generated value out of the spotlight. Out of sight, out of mind.

By keeping everyone in the dark, economists hide viable solutions to serious problems. Spending for Earth tells us about both the business cycle and the economy's surplus. Meanwhile, people needlessly do without crucial information.

While it only takes self-interest to overlook the need to know the worth of Earth in America, it takes social-interest to calculate it. That pretty much leaves the quest for controversial knowledge up to an iconoclast. As usual.

CHAPTER 7

MISSING LINK CAN DE-MYSTIFY ECONOMIES

The "machine" – incredibly complex, inscrutably intertwined – does something absolutely mundane, like making a piece of toast. As it turns out, that's how the world works.

– Adam Felber, Schrödinger's Ball

ECONOMICS AS IF ANSWERS REALLY MATTERED

A team of British economists – Josh Ryan-Collins, Toby Lloyd, and Laurie Macfarlane – co-authored *Rethinking the Economics of Land and Housing*. They ask:

• Why isn't location taught in modern economics?

• What is the relationship between land and the financial system?

• Why are "house" prices rising faster than incomes?

They had to pose the questions because conventional economists don't. How economies actually work differs from how they are currently explained.

Doing economics without land, which with labor and capital is one of the three basic factors in production, is like doing meteorology without water. Lacking land, mainstream economists render the flow of rent into the missing link in economics, a link that explains both fortune and recession.

Since nobody's efforts ever produced land (duh), our spending for land and resources eats at spending on produced goods and services which regularly results in recession. Our spending on the never-produced does not reward anyone for creating the earth (duh-duh).

TREMENDOUS TURF

Our normalcy bias tells us poverty is normal. But how? Bounty overflows from economies. Obviously, poverty is due to concentration of capital. Wrong. Try control of land.

Some almost get it but paint with too broad a stroke:

• the value of real estate is in large part the value of location;

- buildings age, need maintenance – they depreciate;
- what appreciates, despite the regular (and recoverable) down-turns, is the value of the location;

This mistake is common enough. We say a home or housing is spendy while in reality *it's the underlying location*. What makes people want to pay for a site are *factors both natural and social* – the amenities.

- Where land is fertile, views breathtaking, mineral deposits thicker, harbors deeper ...
- Where society has infrastructure, honest citizens, user-friendly government, clean environment ...

There people can earn more, not by working harder, nor by investing smarter (although those two can pay off, too), but merely by operating on a location with such advantages. *Residents and businesses pay more for these advantages due not to owners but to nature and society.*

> *The saying [location, location, location] is repeated three times for emphasis, and it is the number one rule in real estate, though it is often the most overlooked.*
>   — "Why Location Needs to be Repeated Three Times" by Elizabeth Weintraub

As much as one may want to increase the supply of lamd, one can not. What's here now is all that there's ever going to be. As population grows denser, their increased demand bids up what one must pay.

EYES ON THE PRIZE

Measuring rent makes it real, and once real, useful. Tracking changes in the size of our spending on land and resources enables one to forecast coming booms and busts reliably. Already some companies base accurate predictions on land prices. Yet despite their profiting, business lacks the authority of academia, who enjoy authority despite their being unable to predict. Go figure (Ch 28).

Rent-trackers could lay bare the means by which wealth does not trickle down but pours upward. The growing wealth gap is due not so much to the elite behaving badly as it is to everyone behaving normally. We adhere to custom, including counterproductive ones – rights without duties, power sans responsibility. But what's fun about blaming ourselves?

Other factors bolster the prevailing worldview of capital guilty, rent invisible.

- The press reports hi-tech minting billionaires, and previously oil fueling fortunes, but not prime sites enriching a handful of absentee owners and lenders.
- Most people, never buying or selling downtown lots, have no idea how valuable downtown is. On 3-D maps, central city sites tower over others.
- Academia dismisses land as a major factor in generating income.
- And bureaucracy makes a morass of statistics re rent.

However much this surplus, it's due to the occupying populace. As creator of that value, *society has a surplus*. Society could see the world differently – not one of scarcity but one of plenty. Residents might contemplate what's best to do with it. Would people redefine work and play? Feel more worthy of justice? See something so worthy, the Earth, as deserving of being kept healthy?

Besides forecasting reliably and reporting surpluses, economists could offer corrective policies that are both rational and successful. Economists could finally benefit society. Making sense, they'd deserve to be considered authorities.

Authors Ryan-Collins, Lloyd, and Macfarlane reveal the intimate connection between major challenges – including housing crises, financial instability, and growing inequality – and the land economy. Those authors urge politicians and economists to rethink rent.

Presently our BEA, nor any official, furnishes the missing stat that'd plug the gap in economics. If not them, who would? Who's a renegade? Whoever does heed the call to measure all natural assets, they could be in the running for the "alternative Nobel," the Right Livelihood Prize.

CHAPTER 8

THE CALL OF THE RILED

My students say I never listen to them. At least I think that's what they said.
– Prof. Anonymous

WHO CALLS OUT?

Some Earthlings, actually, would welcome knowing a measure for the worth of Earth in America.

• To a curious economist – a "geonomist" – the measure is a piece in the puzzle of doing economics properly.

• To a practical businessman, it's a harbinger; some investors suspect our spending for land drives the business cycle (Ch 28).

• Environmentalists already impute a *replacement cost* for the environment (bizarre, I know, right? see Chapter 18); they'd like to compare that to a concrete figure.

• Supporters of government are always on the alert for a new, better and bigger tax base. They'd like to know how much socially-generated value may be available. And …

• To a fanciful geek, it's a key to conjuring a solution to the inequality problem.

Are you a bureaucrat? Or do you know one? Our species needs the government, an agency, somebody to release the total for the value of land and resources. Maybe you are or know a renegade economist to extrapolate that figure from existing figures?

Once before, fans of economics called for a new stat. Presently GDP measures quantity of growth, not quality of growth (Ch 38). E.g., treating cancer and curing cancer are indistinguishable. Dissatisfied, critics proposed new indicators of wellness and to inform policy. None weighted the current strongest stream within the GDP – rent. They were successful in winning some press but not change. Perhaps a tally for social surplus could satisfy both Old Guard GDP watchers and vanguard livability watchers

Who Has Soundproof Earwax?

You own the land, you own its rent, right? Or maybe not. Still, we grow up thinking of the worth of Earth in America as part of private property, despite society generating the value of location. Want to correct that widespread misunderstanding? Given human instinct, broaching land ownership is like touching the third rail of economics. Rather than call for an answer, propertarians, if they call for anything, it's for silence on this issue.

Heeding their call are the professional statisticians. Those who could, but don't calculate – or don't release – the combined size of all rents, is a pretty long list:

- professional society of appraisers, of assessors, and of bankers (since they're mortgage lenders);

- government agencies who collect the tax on property, both local and state, and apply the legislated exemptions to the property tax;

- agencies commissioned with supplying lawmakers with statistics, both local and state legislatures.

At the federal breadth (not "level," that's too hierarchical for any democracy), the ...

- National Bureau of Economic Research (NBER),

- Census Bureau,

- Departments of Labor, Commerce, Treasury, and Housing and Urban Development (HUD),

- Bureau of Land Management (BLM),

- Federal Communications Commission (FCC),

- Government Accountability Office (GAO),

- Office of Management and Budget (OMB),

- Federal Reserve (not really part of the government but intimately linked to its corridors of power), and ...

- the CIA or NSA. Seriously. The CIA website will tell you how bad the Gini quotient (concentration of ownership) is in the US. What? Spies don't keep stats like that top secret?

All the above alphabet soup – that's a lot of bureaucratic firepower. To not be able to come up with the total of society's spending for land and resources each year – or every day! – is pretty impressive. It must be the single biggest act of bureaucratic neglect since the *US lost some weapons-grade nuclear material* or *lost $6-$20 billion in cash during the Bush-Cheney invasion of Iraq and just looked the other way.*

Public servants are charged with keeping track of public assets. When public researchers don't, and instead advance a distorted worldview, don't you have to draw the line? And when that line is crossed, speak up?

I guess not. Most economists and statisticians sit this one out. They measure wages, profits, unemployment, inflation, stock prices, bond yields, household budgets, the uncountable "consumer confidence" – a myriad of items awash in numbers. But not a total for spending that doesn't reward anyone's effort.

Were they to tackle anything further "out there," they might jeopardize their own mainstream interests. And "out there," if anywhere, is where conventional people place the quest to know Earth's worth. Yet potentially, it's our commonwealth.

WHO'LL TAKE THIS WATCH?

Who could supply us an answer? Assessors compile raw data. Government statisticians access the figures. Economists are proficient at number-crunching. For all three professions, determining society's rent – spending for parcels, fields, and forests plus undersea oil and airwaves – should be a cakewalk. If nobody's measuring the size of pure rent, we can at least measure the size of the problem of trying to calculate rent.

Some do estimate housing value. But that combines house, home, and hearth with location, location, location. They rather unscientifically gloss over the fact that land, unlike labor and capital, is no one's creation while buildings are.

Want to try to pry out an answer? Be sure to put the question – "How much do we spend for natural resources?" – to specialized folk in their technical jargon. Even then, communicating the notion of all rents for natural assets, it's difficult. It reminds one of the old comedy routine of Who's On First? by Abbot and Costello.

> Abbot: *"How much is rent!"*
> Costello: *"How much is rent?"*
> Abbot: *"Yes, it sure is."*

CHAPTER 9

OUR TURN WITH THE TORCH

*Practical men who believe themselves exempt from intellectual influence
are slaves of some defunct economist.*

– John Maynard Keynes

BETTER THE DEVIL WE KNOW – REALLY?

Currently, most of us are stuck in a worldview of the economy as something that happens to us. Despite living amid plenty, we endure dehumanizing poverty, scarcity of free time, overbearing elites running amuck, plus a sickening of the natural world. Then we flail about for solutions.

> *It is hard to imagine [in America] poverty that is worse than this, anywhere in the world. Indeed, it is precisely the cost and difficulty of housing that makes for so much misery for so many Americans, and it is precisely these costs that are missed in the World Bank's global counts.*
> — "The U.S. Can No Longer Hide From Its Deep Poverty Problem" by Angus Deatonjan; *New York Times,* January 24, 2018

Disciplined economists, content with current axioms, continue to tread water. Do we curious ones follow suit or raise the bar and nail down that number for the worth of Earth in America? Such a stat would help show that people respond differently receiving payment for land and resources than they do receiving payment for goods and services.

When hearing of requests to know the value of land and resources, most economists and statisticians seem to recoil in horror. *They want to know how much we spend on the nature we use? Heaven forbid!* Finding that out would capsize the conventional economics boat. That good dunking alone lures me onward.

PROFESSIONALS WONDER ELSEWHERE

Any other economic topic is fair game, but not society's spending for the land and resources they use. Lacking a total for rent, it can't be used in calculations. Without an awareness of rent, it can't be used in theorizing.

Consider the standard explanation of the Great Depression. The supposed cause was the crash of the Stock Market and the purported cure was the entry into World War II. Unmentioned is the huge role that the rollercoaster ride of land values played. Before the crash, *speculators went belly-up in Florida, farmers went bankrupt in Oklahoma.* Turn from cause to cure; before we went to war, the economy was already rebounding. Hence the crash-then-war story amounts to a widely accepted urban myth – and land is a blind spot.

> *Although Oklahoma agriculture had been in the doldrums for a decade, signs of the Great Depression emerged only in 1930 as a drought hit the region.*
>
> – "Great Depression" by William H. Mullins
> at Oklahoma Historical Society

To address economic problems, many laypeople favor one or another conventional proposal. Economists who:

- suggest a stimulus – don't distinguish between spending that rewards effort and spending that does not;

- endorse austerity (for others) – don't calculate the size of society's surplus; and

- extol growth – can't predict the business cycle.

All those competing voices generate much noise for the rent-signal to penetrate.

Most economists consider capital and overlook land, relegating it to a lesser item in the category of tangible assets. In their take on reality, capital accounts for growth. Yet it's the soaring price of land that stymies growth.[1]

Economists who ignore privilege propose what does not work. Complicit politicians, who can't deviate from the norm, adopt what does not work. Officials bark up the wrong tree while society marches along like the apocryphal lemmings off a cliff. All three prove Einstein (a proto- geonomist) correct. Repeating the same behavior while expecting different results is madness.

---

1   "Land-shackled economies: The paradox of soil"; *The Economist*, 4 April 2015

Once a critical mass sees how cost of living depends on location rent, they may demand the statistic from responsible agencies. So far, however, a total of all rents that's serviceable has not dropped into our laps. It's a bit much to expect hirelings to go against the grain, even if public servants, paid with public funds, should serve the public at large.

Yet ascertaining that number would win key insights: we could antici- pate booms and busts, while not knowing the stat loses us that insight. We remain ignorant of the value of the never produced, unable to shake our collective poverty consciousness. And economics will likely stay adrift.

TIME FOR AN ENCORE?

Once found out, the annual total of all rents could be an "ah-hah!" mo- ment for economists. The discipline could embrace the challenge of prediction and develop into an actual science (Ch 28). Economics could play its role of helping make life on Earth as easy and as pleasant as possible.

At last able to watch rent flow, people in general could grasp how econ- omies work, why sometimes they don't, and what we can do about it. We'd figure out how to stop working for the economy and have it work for us, as it is supposed to. We'd get to live in societies with high-performance economies.

Some mainstream economists have felt aroused by the lack of good data. Some have taken a stab at calculating an accurate figure for the worth of Earth in America. Those who went before should not have labored in vain.

Let us gadflies root around, see what those specialists had to say. Googling should dislodge the statistic easily, eh? The keys are the key words. Which will lead to the number for the current worth of Earth?

Even if the web bears fruit, to cover all bases, let's also visit libraries. Hard to believe, but not every printed word is in cyberspace; some still hide away in real books and journals. We'll dig into the US Information Almanac and check card catalogs for tomes never uploaded.

Besides being an armchair investigator, we can trot on over to the hall of records, see what officials have to say. We must pester them, librarians, and researchers. Who knows what overlooked nuggets they may hold?

This could get interesting. Johann Wolfgang von Goethe was para- phrased by an American, John Anster (1835), writing, *"Whatever you can do, or dream you can do, begin it. Boldness has genius, power, and magic in it. Begin it now."* Such perfect inspiration for a Don Quixote. Whatever we find, we'll eagerly report back. Expect to be inundated with data.

CHAPTER 10

DATA – QUANTITY, LOTS; QUALITY, WELL …

Statisticians never have to be right – only close.

A PROFESSIONAL WONDERS WHY

We moderns have lived long enough with our blind spot – land. A good figure for the worth of Earth in America, once known and put into wide circulation, would fill it in. The sum total of the value of assets nobody created has the potential to de-mystify economics for millions.

Yet academics don't research the rents that society pays for all natural resources, including locations, that it uses. Professors don't either. Nor do they expand their work to include all land uses and all non-surface resources.

Karl E. Case, writing for the National Bureau of Economic Research said, *"While a great deal of attention has been paid to house prices in the United States, economists have devoted very little time to the study of land prices. In fact, there are virtually no generally available data on land prices in the United States. It's not clear why this is so…"*

Statisticians and economists explain:

- It's not necessary, the combined values of land and capital work just as well.

- Those two values cannot be separated.

- It'd be too small to bother tabulating.

- Other parts of the economy, like the stock market, are far more important.

- Limited research funds would be better spent elsewhere.

- It's really nobody's business but the landlord's and the lender's.

- Delving into the issue would only fuel class warfare.

Responses like that (Ch 17) provide stiff opposition to anyone seeking information.

Before beginning my research, my interest in the rent phenomenon brought me into contact with economists who did seek this figure. Decades ago, Steve Cord (US; Ch 13), Ronald Banks (UK), Terry Dwyer (Australia) let their curiosity get the best of them, went out on a limb, and estimated a figure for their compatriots' spending on Nature.

Those three lone voices in the wilderness are like sirens – not the kind who'd pull us off course but the kind who paved the way and kept our goal in view. A big thanks to them. Following in their footsteps …

Too Much of a Good Thing?

We have searched the academic databases for every previous article, report, and study for reliable estimates of the value of land and resources. We tried the phrase the "worth of Earth in America." Then "total (or aggregate) return to land." Next, "total (or aggregate) spending for land."

It turns out, public gatekeepers keep nothing from visitors – nothing of value is always available. Want to know the value of all land and natural resources? Good luck laying your hands on it. Bureaucrats don't provide it, at least not in layman language.

Switching to formal language academics use, we typed in "total (or aggregate) value of land" and "total (or aggregate) land value" and "total (or aggregate) rental value of US land" and "total (or aggregate) price of land" and "total (or aggregate) land price" and "total (or aggregate) resource value" and "total (or aggregate) value of resources" and "total (or aggregate) value of natural resource wealth." That swamped us in an outpouring of entries on older scholarly efforts.

Next we limited our dragnet to the last 14 years in order to capture before and after the most recent recession. The list of titles was still long, yet few were current (Chs 13-15). From so much chaff, here's some wheat – guesstimates by specialists:

- At *MSN Money*, "What US Land is Really Worth: Land and Property Values in the U.S." Turns out, rent is very much an urban affair.

- At *New York magazine*, "Because We Wouldn't Trade a Patch of Grass for $528,783,552,000" Can one city park truly sell for so much and the whole country only 12x as much? Hmm.

- At *BizShifts-Trends*, "Imagine the Ultimate Business Mega-Deal: Sell the Planet Earth– All Its Resources, Assets… But; What is the Earth Worth?" Quadrillions. And I thought Central Park cost a lot!

Developed land, or land where housing, roads, and other structures are located, was valued at an estimated $106,000 per acre, while undeveloped land was estimated at $6,500 per acre, and farmland at only $2,000 per acre.

– "What US Land is Really Worth, State by State"
by Thomas C. Frohlich and Alexander Kent;
24/7 Wall St., *MSN Money*, July 3, 2015

Professionals – More Cautious Than Curious

To answer Karl Case, researching how much we spend on the Nature we use puts a bullseye on "passive income" politely, on "unearned wealth" pointedly. Does that discomfit some? With two-thirds of Americans banking on their houses, we're all speculators now – or dream of becoming one (Ch 25).

Behind the scenes, is there pressure to toe a line? Is an inquiry into the value of land the third rail of economics? As is Proposition 13 – limiting property taxes in California – the third rail of politics? Touch it at your peril.

The bias against treating rent scientifically has been internalized, like QWERTY on the keyboard has been institutionalized. Without being told, economists know what's considered a legitimate area of research and what's not, as do their students. Then comes the preponderance of courses and seminars and papers on everything but land. Few personality types can buck the trend. Most steer clear of the bigger picture figure. (Chs 6 & 12)

Yet shouldn't how much the populace spends on its natural heritage be pubic knowledge? Should public servants serve not an insider power but the general public? They'll have to eventually. Economists cannot stay unscientific forever; someday they'll need that stat. Truth will out, won't it?

Our Holy Grail – that numbered needle in this information haystack – will be located. We located those who take stabs at knowing part of Earth's worth. We'll begin with the articles written for popular consumption.

CHAPTER 11

MEDIA REPORT OFFICIAL LAND VALUES

The mass media offer entertainment to be consumed, forgotten, and re-placed by a new dish.

– W. H. Auden

THEIR NEW DISH

Cutting-edge bloggers in recent years have published what they have discovered about how much we pay for "Earth in America." Reading their articles in the popular press let us skirt the jargon. We may even find a shortcut to the value of land and resources. But it'd be accurate only if these reporters don't just report what's been written. They must also weigh the researcher's methodology for thoroughness.

TOP 10 ARTICLES

1. 2009 May 24, "Total Assets of the US Economy" at the blog of John Rutledge:

"Why is it that people know so much about something so small (GDP) but so little about something so big (total assets)? A balance sheet for the US economy at the end of 2008 shows (in billions) $104,049 in total assets divided between $58,639 in financial assets and $46,301 in tangible assets. People own more than $46 trillion of real stuff like land, buildings, cars, and computers, not just securities. The Federal Reserve does not report complete balance sheets for other sectors (farms, financial sectors, federal government, state & local governments) or rest of world (foreign owners). I think that is a big mistake. The other sectors are actually bigger than the ones they do report. Adding in farms, financials, governments, and foreign owners brings the total financial asset figure up to $141,512 billion at the start of the recession. *As of June 2013 the market value of total assets held in the US rose to $234 trillion, not counting the value of land and natural resources held by governments (the Federal government owns more than 700 million acres), corporations, or foreign persons."*

Consider John's numbers. Of total assets, approaching half is tangible assets. Of tangible assets, well over half is real estate, likely over 80%; land plus buildings are worth one heck of a lot more than cars, computers, clothes. And half of real estate is land. How much? Using John's ratio and his $234t, tangibles would be about $90t, real estate about $72t. If land is one half and buildings the other, land's aggregate price is $36t – way up there, yet based on Fed figures. (In Ch 13 we reconcile that.)

2.   2012 July 13, "The World's Most Resource-Rich Countries" at *Business Insider* by 24/7 Wall St:

"Using the most recent uniformly available data on reserves and global market prices, 24/7 Wall St. calculated the total value of the proven reserves of 10 of the most valuable resources, by country. They include oil, natural gas, coal, timber, gold, silver, copper, uranium, iron ore, and phosphate. The U.S.'s total combined resource value [mostly fossil fuels and forests] is $45 trillion. One could make a good argument that as a filter for carbon dioxide emissions and as a producer of fresh water, the world's timber lands are even more valuable untouched than they are when cut down and sold."

Their $45t price tag for US resources dwarfs Rutledge's figure for land – likely $36t – and does not even include ordinary surface use like housing. (In Ch 15 we reconcile that.)

A slew of articles followed the next year in 2013:

3.   July 8, "What's Land Worth?" in *BEEF Daily* by Amanda Radke:

She notes how ranches – called *"a safe-deposit box with a view"* – now sell mostly not to other ranchers but to urban investors and outdoorsmen who are willing and able to meet prices that are beyond what many cattlemen can afford.

Whether rural or urban, if desirable, people with cash-to-burn bid up land's price. Land tends to hold, even swell, its value – except when it doesn't. Yet the downturns don't last as long as the up. And when prices swing up, the upswings always bring a new top.

4.   Sept 3, "A Remark on Ricardian Taxes" at the blog of Ashok, a college kid at the University of Pennsylvania (who was cited in the next article). The cited David Ricardo was an early economist (and successful stock trader) who showed land does not drive the price of corn but rather the value of the output drives the value of the location.

Ashok projected:

> "... *minerals can be taxed under Ricardian principles, there is a gold mine of untapped tax revenue. Just look at the sales of America's oil and gas sector* [as did 24/7 in #2 above]. *Along with land, and maybe a small levy on income earned above $1M we can finance a big government (25% G/GDP) with* sensible *tax policies."*

Tax millionaires? If government collects the annual rental value of things like minerals, oil, and land, how many incomes could still exceed $1 million annually? Enough to bother about taxing?

That aside, what does he mean by "big government." Like many people, probably he means *federal* government. But why shoot for big government and not for sensible government, especially if one's touting *sensible* tax policy? I suppose when students eschew tabulating and resort to projecting, they lapse into dreaming.

5.   Nov 22, "End the 1 percent's free ride: Taxing land would solve America's biggest problems" in *Salon* by Jesse Myerson of Occupy Wall Street. Each month, 20 million visitors read *Salon's* award-winning content:

> Besides citing Ashok above, Jesse also cites a scholar who used to work for Chase Manhattan and has done much enlightening research into banking: *"Michael Hudson has assessed that the land value of New York City alone exceeds that of all of the plant and equipment in the entire country, combined."*
>
> While that comparison is nice, still, what do we multiply it by to get the land value of the whole nation? And would Hudson agree NYC could sell Central Park for over a half trillion dollars (Ch 10)? Jesse continues ...
>
> "...the amount of revenue that can be raised by taxing the land is huge. Enough, for example, to support truly liberatory social spending, like a universal basic income, without risking inflation."
>
> Our question re the size of Earth's worth in America sure does arouse the interest of those who favor government spending. Perhaps they'll add their voices to a call for officials to tabulate an accurate statistic of society's surplus. When a public agency does respond with reliable numbers, my work here will be done.

In Riverside County California, a decrease of 10% in the distance to the nearest stand of oak trees resulted in an increase of $4 million in total home value and an increase of $16 million in total land value in the community." –2018 Jan 5, "Big Trees Make Your Property Value Grow" at BrightView by Tree Care. Public amenities make locations more valuable. As such, land value is a social surplus.

Later in 2013, *Slate* – which has won numerous awards, including the National Magazine Award for General Excellence Online – ran two articles in their MoneyBox section, one by Ashok (above) and one ...

6.   Dec 20, "What's All the Land in America Worth?" by Matthew Yglesias:

> Based on the Federal Reserve's Flow of Funds report, his total price was $14.488 trillion. However, watch out; Matt passed on the academic's inflating the price of buildings which deflates the price of land. Maybe that's why this total is way under Rutledge's likely $36t. It also seems to measure only the surface, not resources which at $45 trillion are three times $14.5t.

7.   Next year, 2014 Feb 6, "Response to Jesse Myerson's Land Tax Idea" at *The Tax Policy Blog* by Joseph Henchman of the Tax Foundation:

> His back-of-an-envelope total price was $50 trillion, but that included buildings with land. Land alone, he did not say. We'll be fair and guess half, or $25t. That splits the difference between Matt's low $14.5t and John's likely $36t high.
>
> One year later, William Larson at the Bureau of Economic Analysis estimated that the 1.89 billion acres of the 48 contiguous states and the District of Columbia (not buildings, roads, or other improvements, nor bodies of water) may sell for $23 trillion (Ch 13).
>
> That's close to Joe H's likely $25t but far from John R's likely $36t. In Larson's wake came another batch of articles, in 2015:

8.   Apr 22, "How Much Is the U.S. Worth?" in *Wall Street Journal* by Eric Morath:

> "Land has long been recognized as a primary input in production and as a store of wealth. Despite its fundamental role in nearly all economic activity, there is no current and complete estimate of the value of the land area of the United States."
>
> Tell me about it.

9.   Apr 23, "Fun Number: The US Is Worth $23 Trillion. Or, Why A Land Value Tax Won't Work" in *Forbes* by Tim Worstall:

> Tim points out 5% has been the average rent on land for long historical periods in a number of different places. So, 1/20th of Larson's total yields $1.15 trillion rent per year from unimproved land – not much if you compare it to other income sources or to government spending. We'll keep looking and find out if 5% still applies and how much more all kinds of land would bring in.

10.   Jun 7, "How much would it cost to lease all of the land in the United States per year?" at *Quora* by Riley Ashton (until, for some reason, taken down):

> Riley found subtotals for private land, both residential and commercial, and for public land, but only federal, and listed the sources. That leaves out some acreage and other uses yet it totaled $2.72t annually (not price, rent). That's over double the estimate of Worstall.
>
> These last two estimates, by Worstall and Ashton, turn from lump-sum price to annual rent. Both of these thinkers are on the right path to a handier number, much more realistic. There is no market of buyers at stratospheric prices but there could be a market of renters at manageable amounts.
>
> The two who want to redirect land value to public purposes, or to pay citizens a dividend—Ashok and Myers—or anyone, be aware. Taxing all the lump sum price would in practice be confiscating the land. The only amount available to taxers that'd not confiscate land or depress its value is its annual rent.
>
> What might aggregate land value mean to you, dear reader? How much is society's spending for nature per capita? Divide registered voters in the US into $3t – a figure including what Ashton left out. It's $12,000 each year, or one thousand bucks every month, per voter.

HELP IS ON THE WAY

I'm excited to have dived in. It feels good to find some sort of answers and rub intellectual elbows with those who went before. But I'm also wary, a bit. While grateful to those who tackled this quest, one must admit most articles were not overly analytical, nor very critical. Sure, one must respect authority, but an investigative journalist must do more than that – s/he must ask the hard questions, too.

Nothing for it but to tackle the research papers, academic jargon and all. Make sure the value for all utilizations of land and resources gets counted, not just surface land for housing in particular or for buildings in general. You know, count oil, water, airwaves, et al. And whoever's doing the counting, weigh their methodology for bias toward too little or too much rather than the Goldilocks means of just right.

Better input is needed. After refereeing the academic articles, we peel off that layer and drill down to their sources. We squeeze dry official sources and business sources of their raw data.

When questions arise, we call, chat, and email the people who work in the discipline. Once contacting them and putting our queries in the requisite academic jargon, academics aid us by explaining any murky points. Further, researchers tell us:

- if they have more recent figures,

- how big a ballpark, how big a range their estimate fell in,

- why they focused on what they did, why they did not go broader or deeper, and

- who else they recommend to talk to, other sources to check out.

They call academia "the cloister" for a reason, but we'll crack that nut – and the powers that created it.

CHAPTER 12

MONEY TALKS ... AND SILENCES?

You know unemployment has gotten bad when the 1% starts laying off Congressmen.

"RENTENTION" – HARD HABIT TO BREAK

L and is still a goldmine. In "The Rate of Return on Everything, 1870-2015," economists Oscar Jorda, Katharina Knoll, Dmitry Kuvshinov, Moritz Schularick, and Alan Taylor compare stocks, bonds and housing over the past century and a half. They find the return on residential real estate [sitting on land] has been as high as or higher than the return on equity. As modern economies have grown and developed, owners of the ground on which we live have been steadily enriched.

While our seeking to know the worth of Earth in America is an economic quest, it unavoidably runs up against political hurdles, more formidable than many imagine. *One hundred of the wealtiest American families own as much land as New England.* Their shell companies buy up vacant lots in pricey cities like New York. Those banking the payments we make for the nature we use likely intend to keep on doing so.

The wealthy may worry that if public awareness of natural rents were to increase, then their private concentration of such rents might decrease. Upon learning of the immensity of this socially generated surplus, some people would reflect upon the facts that nobody made Earth, everybody needs land, and population density creates its value. Location rent is a free lunch for somebody. As Rex Nutting of CBS MarketWatch notes, "Billionaires haven't earned all they have."

People could consider what to do with all that spending that never rewards anyone supplying any labor or capital but mostly those controlling land. The electorate might choose a more equitable arrangement ("The time may be right for land-value taxes," *The Economist*, 9

August 2018). Inequality is a concern of even the market-oriented (and of a fellow Oregonian).

> *"A 1% increase in housing costs [housing on land] increases income inequality by 0.125%, spending inequality by 0.248%... [in a] survey of 1,767 leaders from academia, business, government and non-profits, The World Economic Forum ... found increasing income inequality to be top global concern in 2015."*
>
> *– The Conversation*

If you're accustomed to receiving rent, what are you going to do? Forgo it? Or spend some of it to keep on getting most of it? Those who now receive the lion's share of our spending for land – wealthy lenders and speculators – don't lack for power and usually get what they want. Have they shaped data gathering? Do rentiers discourage research into land and rent? So other governments don't take an Aspen turn? (Ch 40) As a who-done-it, the obfuscation of rent is an open-and-shut case. The motive is the possibility of losing rent; the means is rent itself.

Forbidden Knowledge

That the rich and powerful pay the piper and call the tune is par for the course. Suppressing research, monopolizing and massaging knowledge, it's what elites do. Centuries ago, if an astronomer measured planetary orbits exactly, the ruling elite had kindling piled around a stake. While the *auto da fe* is not currently a threat, the Goliaths (veteran distorters of truth) still keep a lid on the Davids (wannabe discovers of truth).

While the gentry usually enjoy smooth sailing, occasionally they encounter stormy waters. A century and half ago, the biggest fortunes were amassed by those able to win ownership of railroads – which actually raked in more money from the land they were given by Congress than from the freight and passengers they carried. Congress, in exchange for stock and seats on boards of directors, also granted near monopolistic ownership of timber for track ties to what became today's Weyerhauser.

The people did fight back with muckrakers and unions. Also, millions from all walks of life, inspired by Henry George during the final quarter of the 1800s, promoted a *"Single Tax on locations."* In response, John D. Rockefeller *gave $1,000* to Republican candidate Teddy Roosevelt to defeat Henry George, and the powers-that-be robbed George of his victory in the mayoral race of New York.

The third most popular American after Mark Twain and Tom Edison, George was in demand as a speaker all over the world. While on tour in the UK, he had to flee for his life from thugs sent to kill him. What ended his life prematurely was his arduous travel itinerary and too many cigars (one was named after him before the Baby Ruth candy bar was named after the baseball hero), but his followers kept battling.

When the popular movement to publicly recover socially-generated rents was a force to be reckoned with, the rentiers did reckon with it. They quashed the movement at every turn. Since power corrupts, the gentry corrupted not just politics but also economics. Of all the social studies, none is as relevant to the most powerful elements in society as is economics.

Dr. Mason Gaffney, UC-Riverside, tells us in *The Corruption of Economics* that those corporations – AKA the "Robber Barons" – funded the universities that then were creating the nation's first departments of economics.

- Rockefeller oil money at Chicago;

- Timber money at Cornell;

- Railroad money – Union Pacific is still the *biggest* landowner in California – at Stanford and Johns Hopkins;

And so forth.

During the formative stages of the discipline, the newly degreed economists ignored the role of rent and submerged nature's land into humanity's capital. They did so despite land being one of the three factors of production in classical economics, and still is in real-world economies. Those founders focused overwhelmingly on labor and capital, reinforcing the boss-versus-worker paradigm. With land out of the picture, rent got the boot next. There was no reason to count that which no longer counted. As land disappeared into capital, rent disappeared into the ether.

Knowledge Appeals to Privilege

The rich still fund only what they like.

Donors often give such large amounts after they have developed 'long and careful' relationships with universities, and after college officials have a full understanding of the donor's passions.
> – Kellie Woodhouse, "Does Harvard Need Your Money?" Inside Higher Ed

Wealthy donors have strong feelings about how nonprofits should utilize their contributions.
> – Claire Costello, National Philanthropic Practice Executive for US Trust, "Understanding motivations, challenges and expectations of wealthy donors" in Giving Tuesday, owned by Bank of America.

The fraction of time I go ahead and do what he asks isn't as high as you think.
> – U of Michigan President Mark Schlissel said of billionaire real estate developer Stephen M. Ross, who pledged $328 million to the university "Demanding Donors to Colleges, Universities Hold Sway" in Free Press

... 26% of funds donated to universities in 2014 went to endowments with restrictions (meaning income from these endowment gifts is designated by donors to a particular purpose), while 1.6% went to endowments without restrictions on the use of income. Fifty percent of gifts went to restricted current operations, and 7% were completely unrestricted; 13.5% of donations came in the form of or were for the purchase of property, buildings and equipment ...
> – "11 Huge Gifts Made to Universities by the Super-Rich Over the Past Decade"

Major donors – rentiers – dominate the conversation by funding only the research that agrees with them. They ignore the topics that don't suit, or may even threaten, their interest. By determining what's considered legitimate topics of research, they marginalize and even make controversial any formal study of certain fruitful subjects.

For their role, universities are part of the local growth machine via their construction campaigns. They are often some of the biggest landlords in some of the poorest neighborhoods. One egregious example is Yale acting as a slumlord in New Haven Connecticut although they've been trying to do better.

New buildings are needed in part to house the study of not socially-generated rent but of real estate. The number of real estate schools and departments has proliferated at dozens of universities, all funded by the industry.

The accumulated effect of the preponderance of all this funding has been to keep rent, society's surplus, society's spending for land and resources, out of academia.

COMPLIANT ECONOMISTS

Rentiers could lash back at specialists if they go against prevailing winds and reveal the size of rent to the public. But the gentry don't have to. Once a worldview gets institutionalized, it gets awfully hard to change. Normalcy bias sees to that. Outside a rare course-correction, more interference from above is not much needed. Donors don't have to be heavy-handed. They can be quite discrete. They don't censor. They don't have to.

In service of the elite for over a century, the economics discipline has not been self- correcting. Not many if any in the field object to the loss of land as a factor. Thus the lack of scientific rigor in economics is not sloppy or accidental; it's political. It's an inevitable result of subtracting land.

The behavior of academics is natural enough. It's the nature of domesticated animals, such as civilized humans, to obey, even lionize, those higher up in the hierarchy – and donors outrank economists by a long shot. Beyond that instinct, economists are subjected to another one: self-preservation. That gives academics and bureaucrats a more pressing reason not to rock the boat.

By making it difficult for economists to take in the entire panorama of economies in action, the elite have made it difficult for anyone to poke their nose into society's spending for the nature it uses. Lost to sight were not only rents, the sources of great fortunes; Didier Jacobs of the Center for Popular Economics calculated that, "when it comes to the very richest Americans (Forbes' billionaires), 74% of their wealth is derived from rents." (in "Are Billionaires Fat Cats Or Deserving Entrepreneurs?," March 2, 2016). But also lost, to society's detriment, is the driver behind the business cycle. By handicapping economists, rendering them unable to forecast booms and busts, rentiers deprived society of what could have been a useful science.

If you're not a science but want to look like one, what are you going to do? Rentiers control rewards, both monetary and prestigious. They could pirate a prestigious prize and gain the patina of science.

COUNTERFEIT NOBEL

In the perilous 1960s (perilous to the gentry), central banks (owned by rich families of old money) lobbied the Nobel committee to give a prize to economists that the bankers offered to fund. Alfred himself left no money for economics, a field held in such low esteem back then by real scientists. Nobel also snubbed mathematics, some say because a woman he was enamored with was wooed away by a mathematician. Rather than lobby the Nobel committee after Alfred's death to be laureated, the mathematicians created their own prize, the Field Medallion. Yet few have heard of the Field Medal while the whole world knows about Nobel laurels.

Bankers wanted that name's prestige for their pet field and, notes *The New Yorker's* economics reporter, John Cassidy, ponied up the prize money (Dec 2, 1996). The Nobel committee caved, mostly. Rather than bestow the faux prize in the annual awards ceremony in Nobel's native Sweden, economists must do so in nearby Oslo. There, too, do-gooders for peace give out their prize using the family name, beginning well after Alfred died. *Alfred's descendants have asked the bankers funding the false prize to quit using their family name*, notes Hazel Henderson at her site, Dec 30, 2004. So far, the bankers and the committee have turned a deaf ear.

Note the double standard. When Levi's complains about a fly-by-night clothier in, say, Vietnam, slapping the Levi's label on subpar jeans, everyone agrees those jeans are counterfeit. But when central bankers slap the Nobel family name on their favorite economist *du jour,* nobody in the mainstream media utters a peep.

According to Avner Offer and Gabriel Söderberg, authors of *The Nobel Factor: The Prize in Economics, Social Democracy, and the Market Turn,* the award serves elite interests. Global bankers have given their prize to an academic focused on society only once, to the Swede Gunnar Myrdal in 1974 (who, ironically, later turned against socialism). Every other year it went to economists more friendly to business interests.

Many people assume that what the prize covers is science and what it leaves out is not – a huge bias in favor of the status quo and against geonomics. Yet neither faction of economists – neither those who are pro business nor those who are pro society – is very scientific, since neither distinguishes between spending that rewards lobbying for privilege and spending that rewards production of real goods and services. The award has continually reinforced the primacy of the present biased market in which the winning of rent is no different from profiting from production.

Banks, The Collector for Rentiers

Our overlooked land and our conspicuous money meet is at the bank. Banks make most of their money off mortgages. Consumer lending makes up the bulk of North American bank lending, and of this, residential mortgages make up by far the largest share. It's mainly mortgages – payments for land besides buildings – that make F.I.R.E. (Finance, Insurance, & Real Estate) the biggest sector in the GDP.

Locations are valuable and often change hands (thanks in part to job transfers and to the high divorce rate). The turnover of that asset means more profitable business for banks. The higher the bids for land, and the more often land sells, the more banks profit.

According to professors Jordà (also with the Federal Reserve), Taylor, and Schularick (above) in their *The Great Mortgaging* (CEPR's VOX, 12 Oct 2014), banking today consists primarily of channeling savings back to families to buy real estate. In advanced economies, banks resemble real estate funds: borrowing (short) from depositors and capital markets to invest (long) in assets linked to real estate.

They continue. Since the founding of the Federal Reserve in 1913, nearly all of the increase in the size of the financial sectors in Western economies stems from a boom in mortgage lending to households. By contrast, financing the business sector has remained stable over the 20th century in relation to GDP.

Household mortgage debt has risen faster than asset values, resulting in record-high leverage ratios. Mortgage credit has generated financial fragility of household balance sheets and the financial system itself. Contemporary *business cycles are shaped by the dynamics of mortgage credit*, with non-mortgage lending playing only a minor role.

For all this largely land-business to happen, there needs to be plenty of cash available. Bankers merrily make it so. During every period – about 18 years – of the business cycle, they gradually lower requirements for borrowing (see Ch 28).

It's become trite to note how, last decade, bankers dove into subprime lending. What's not reported is how routine that is – they do something similar every cycle. And let's not forget, they over-extend not just to starry-eyed home buyers but to the rich and powerful, too; real estate mogul turned anomalous President, "The Donald," *has been bankrupt six times*. When bankers do this loosening of credit requirements, they may enrich themselves but at the expense of others; inflating land values makes crashes inevitable.

Further, banks gain every planting season when farmers borrow to buy seed, etc. Additionally, banks take property (half of which is location) as collateral, then later for themselves in foreclosure. Banks also profit when consumers fall behind in their credit cards.

Banks gain again whenever a government creates bonds. As the middleman selling them—e.g., Goldman Sachs— to other rich people and large institutions, brokers reap a tidy bit of income. The deeper governments go into enormous debt, the richer such brokers become. For them, nothing's wrong with that picture. The federal government does not have to let broadcasters use the public airwaves for free, or timber companies log public forests and hard rock companies mine public land almost for free, or ground traffickers snap up the land values at the exits off the interstate highway system, or ... You get the picture. It's a long list.

BANKS – OWNERS AND OWNEES

Who are those on the receiving end of society's spending for assets never produced by anyone? A lot of people get a little but a few get a lot. Much of our spending for ... goes ...

- oil to the Rockefellers, Mellons (Gulf Oil), and Kochs;

- food in its raw state to Cargill and Kochs, further concentrating already concentrated farmland;

- the airwaves (telecommunications) to Rupert Murdoch (Fox/News Corp), Sumner Redstone (ABC/Viacom), and Brian Roberts (NBC/Comcast);

- land, typically as a mortgage, such debt being the "RE" in "F.I.R.E., to owners of major banks.

Rentiers have magnified their income by capturing not just rent but also interest on debt. The "oiligarchy" used oil rents to found banks: Rockefellers took over Chase, the Mellons launched Hannover's which now is Citi. Warren Buffett bought the biggest block of stock in Bank of America and Wells Fargo.

It's not just for profit that the already rich go into banking but also to control the spigot of credit and the issuance of money. Since winning that power from Congress over a century ago, despite the Constitution making Congress responsible for that function, bankers have printed dollars and set interest rates, once pegging it at 18% (1981). Headed the other way, notes the Fed's St Louis branch, has been the purchasing power of the dollar: it has shrunk while federal government debt has swollen.

Inflation magnifies a bank's income, and more so for a big bank than a small one even at the same rate of inflation (Ch 26). Banks, as do other big businesses, tend toward giantism. Big banks gobble up little ones so much that now five banks hold almost half of the assets held by all banks. With inflation, recession helps banks, too, whose assets have quadrupled since the most recent major downturn.

Banks are the most common type of organization that controls the shares of the 299 biggest global corporations. Financial institutions – banks, financial companies, insurance companies, or mutual and pension funds or trusts – owned the majority (68.4%) of shares in these dominant corporations. Financiers control the corporations. And while it's profitable, stable, and influential to control corporations, don't lose sight of the fact that banks make most of their money from mortgages. Controlling both land and money, what more could you want?

BANKS – OWNEES AND OWNERS

BlackRock, founded and operated by the current generation of Wall Streeters, held or controlled 6.1% of the assets of companies. At the end of 2017, they managed real estate, stocks, and other claims priced at over $6 trillion. BlackRock and Capital Group, which grew up with Los Angeles (not New York), have both wide influence across many companies and deep influence, often being the top or the second-ranked shareholder.

About 1.5% of shareholders of those 299 companies owned or controlled some 51.4% of the assets. These largest shareholders numbered only 30 out of more than 2,100. They were made up of 21 private-sector entities and nine public-sector (that is, government-owned).

In general, the tendency of this economy to concentrate wealth, based on society's current definition of property, continues apace. Out of over 3,000 US corporations, only *30 of them rake in half of all net corporate profit*. That's 1% ... again.

US-based mega-corporations accounted for 29% of the 299 companies. Six of the top ten private shareholders were based in, or at least originated from, the US. So did ten of the top 21. The top eight shareholders each held shares in more than half of the top 299 corporations. Eighteen of the top 21 shareholders each held shares in at least 100 very large corporations.

BlackRock, Vanguard, and State Street are the biggest owners of the 4th biggest bank, CitiGroup. Vanguard, BlackRock, and other index funds own about 30% of REITs (real estate investment trusts). REITs are the tax-protected investor funds that own many of the nation's offices, shopping centers, apartments, hotels, warehouses, data and storage centers, and of course their very valuable underlying locations; they've parked $1 trillion of their eleven trillion in REITs. Vanguard is by far the largest owner of REITs.

After the housing bubble popped, millions of Americans lost their homes. Private-equity firms led by Blackstone quickly bought tens of thousands of homes at deep discounts, most of them out of foreclosure. Blackstone – which spun off BlackRock – became the largest single-family home landlord in the US, with 50,000 properties.

BlackRock, Vanguard, and State Street have nearly $11 trillion in assets under their management. These Big Three have become the largest shareholder in 40% of all publicly listed firms in the US. They are the largest single shareholder in almost 90% of S&P 500 firms, including Apple, Microsoft, ExxonMobil, General Electric, and Coca- Cola. The Big Three exert the voting rights attached to these shares. They vote for management in about 90% of all votes at annual general meetings, while mostly voting against shareholders proposals (such as calls for independent board chairmen).

Who owns The Big Three? All three own each other. Big chunks are also owned by big banks. Which real persons actually own them, is hard to find out, but it's undoubtedly the 1%. The true ownership of shares is hidden by the use of "nominee" or "depository" organizations – such as the Depository Trust Corporation in the US.

Even if their first fortune was not made in land, they soon invest in land. Zuckerman, Gates, *et al* own entire islands. Gates owns thousands of acres in Arizona, Florida, and Pennsylvania. Lesser rich buy up farmland as a safety deposit box. The Dreyfuss Fund (the family of actress Julia) owns prime downtown locations in most major American cities, along with other family funds. A very few families own very much prime land.

Owning the best is not the only thing the rich have in common. The ties that bind are their marriages, old school ties, nepotism, interlocking directorships, interlocking stock ownership, ties to politicians, offices held, etc. The super rich meet at hideouts like the Bohemian Grove. The Rockefellers' Trilateral Commission seems to have decided the world's fate. While their sources of great fortunes vary, on key issues they present a united front. Who dug this up? G. William Domhoff of UC Santa Cruz.

Concentration is hard to exaggerate. This tired democracy is really an aristocracy. All based on corralling the torrential stream spent by humanity on our need to use land.

The Real State is Real Estate

One intractable and widespread blind spot is the popular belief that the state and the elite are separate entities. Only on the surface. And even there, the elite and state cell-divided only a few centuries ago when aristocracy and government split rather amicably and monarchs made space for parliaments then consisting of nobility. The slogan of *"government of the people, by the people, for the people"* is quite modern and true more in word than deed. Who gets laws passed? Who gets to violate them with impunity? Who gets bailed out? Who gets a foreign policy serving their interests? Who contributes the most to political campaigns, besides to foundations and universities?

When subsidizing the Nobel committee or endowing an Ivy League university, how do the rich deliver that fat check? Face to face? Electronically? Via a minion in an office? In person at a fancy dinner party? What are the intimate details of how the real financial world works? The truly aristocratic rich do not have jobs; they have people for that.

Who outranks whom? Whoever pays the most. Endowed with a torrential income of rents, the landed elite contribute mightily to lobbying and political campaigns.

Between 2007 and 2012, 200 of America's most politically active corporations spent a combined $5.8 billion on federal lobbying and campaign contributions.

Wealthy donors, comprising less than .01 percent of the population, accounted for 40% of all political contributions in 2012. All those donors could fit in a baseball stadium. Over 93% of SuperPAC money came from just 3,318 donors (that'd be a minor league park); 59% of it came from just 159 people (that'd be a public park).

Policies supported by economic elites became law 60% to 70% of the time. Policies supported by business lobbies became law 60% to 70% of the time. (Often these were the same policies.) Policies favored by a majority became law 30% of the time and only if the economic elite or business lobbies (or both) also supported them. *"The opinions of the bottom 90% of income earners have a statistically non-significant impact."* (in *Minn-Post* by Eric Black, 08 May 2015.)

Of the 200 donor companies analyzed for *Fixed Fortunes*:

- 7 are in transportation, which includes road-building – and that takes valuable land;

- 11 are in oil and ores, etc, whose profits largely are rents;

- 13 do weaponry, used to control others' resources and sea lanes;

- 13 do agribusinesses, obviously on farmland;

- 21 do healthcare, whose high cost is due in part to ruining land and environment;

- 28 do communications and electronics, which includes some utilities; and, ta-da,

- 48 are in finance, insurance and real estate. F.I.R.E. is consistently the largest source of campaign funds for federal politicians cycle after cycle. One payoff was the bailout of trillions and trillions of dollars.

For their contributions, the 200 corporations received 760 times as much subsidy – $4.4 trillion in federal business and support. That's more than the $4.3 trillion the federal government paid the nation's 50 million Social Security recipients over the same period. *The $4.4 trillion total represents two-thirds of the $6.5 trillion that individual taxpayers paid into the federal treasury.*

Over 85% of the 200 corporations won subsidies from state and local governments, too.

The industries who spent the most to milk the US over 10 years are – in billions:

- Agribusiness, which takes place on farmland: $1.21;

- Weapons, which defends rentiers: $1.26;

- Drug companies, whose customers often suffer from land abuse: $2.16;

- Fossil fuels, a kind of "land": $2.93;

- Communications, utilizing utilities in land: $3.50; and the usual kingpin,

- Finance, or mortgages, basically: $4.29.

In the latest ranking, F.I.R.E. still comes in first, by a long shot. If you add natural resources and farmland, plus the rent components of all the other sectors, it's no contest. Rentiers dominate overwhelmingly.

No other component of the ruling elite donates as much to electoral campaigns as do the recipients of ground rent. In local election campaigns, the real estate lobby contributes the most. In federal election campaigns, F.I.R.E. contribute the most.

To paraphrase Freud, what do the wealthy want? The usual. A hunky subsidy. A sweet-talking tax break. A party that fawns all over them. And without having to publicly, explicitly reveal their darkest desires – "stealth politics." They want to be understood without having to be explicit. If we're intimate, it goes without saying; you get it, right?

SILENCE IS GOLD-RIDDEN – THE SPOILS OF THE SPOILED

The holders of public office determine public budgets, which funds the bureaucracies who hire statisticians. So far, no one in authority has put the tabulating of rents in any official's job description. So there is no official figure. The bureaucrats who collect statistics have neglected rent for as long as they have been collecting statistics – and likely will until economists realize the key role of land and request a measurement of rents from officialdom, regardless of whose feathers would be ruffled.

The gentry contribute mightily to politicians and in exchange receive mighty favors worth beaucoup billions. Routine favors include subsidies. In bedroom communities, Eben Fodor, author of Better Not Bigger, figures that each new house costs the taxpayers $25,000 to finance the new roads, sewers, schools, etc that the homebuyer did not pay for – a giving, just the opposite of a taking that land hoarders complain about. And downtown, landowners do not cover the cost of improvements to the infrastructure, or compensate neighbors for the loss of direct sunlight, etc.

Instead of paying taxes, the gentry win tax loopholes – deductions, depreciations, deferments, exemptions, etc. Such tax breaks not only enable the favored donors to keep rents rather than pay them to the public treasury, they also inflate the price of locations, further enriching sellers, developers, and lenders.

Some call those tax breaks, "tax subsidies." Over half of them go to F.I.R.E., utilities, broadcast spectrum, and Big Oil (Joe Romm for Think Progress, November 13, 2011). Agri-business gets huge actual subsidies plus is excused the cost of their toxic chemicals contaminating others downstream and downwind. Mining corporations, the same story. Loggers, same thing.

When commerce slows and threatens big business with bankruptcy, politicians favor insiders with real subsidies, the best they can offer: bailouts. After the last recession, both the US Treasury and the US Federal

Reserve pitched in with almost \$30 trillion.[1] The Fed rescued not just banks but also other Big Business, not just domestic firms but foreign ones as well, in which the elite invest.

Turning from legislatures to the executive branch, those agencies charged with enforcing the law do so selectively. Oil companies, for example, often fail to pay royalties without being punished, and pollute the environment egregiously without being fined. Locally, it's meat packers. Industry in general gets a pass.

Harvard professor David Landes and other economic historians say the key to great fortune is to socialize your private costs—like pollution—while you privatize everyone's social gains—like the value of locations. Turning from the executive branch to the judicial, the first Chief Justice of the US Supreme Court, John Jay, said, *"Those who own the country ought to govern it."* The court has never wavered from that founding mission. Some of the most famous decisions by the Supreme Court favored land speculators. Justices ruled ...

- 1785, land contracts based on fraud are valid;

- 1880s, railroads do not have to return unused land to Congress;

- 1880s, railroad land is tax-exempt from any jurisdiction;

- 1895, after progressives shifted taxes from tariffs to income – largely the rent for land held by the top 10% – tariffs may shrivel but taxes on rents may not stay;

- 1900s, tabled cases filed by the first Chief Forester Gifford Pinchot, since they could have set off enough litigation to throw the entire Western U.S. into turmoil;

- 1987, *Nollan v. California Coastal Commission,* sided with a developer to block an ocean view;

- 1992, *Nordlinger v. Hahn,* upheld the constitutionality of California's Proposition 13; and ...

- 1994, *Dolan v. Tigard,* sided with an owner to build on a riverbank;

Despite their hegemony, the gentry remain vigilant to any potential threats. When the Soviet Union collapsed, the cutting-edge Soviet economists invited some big name American economists to come over and tell them all about non-left, non-right, third-way economics, centered on

---

1   "\$29,000,000,000,000: A Detailed Look at the Fed's Bailout by Funding Facility and Recipient" by James Andrew Felkerson of Levy Inst. at Bard, WP 698, December 2011

a proper social role for the flow of rents. Fred Harrison wrote in *Land & Liberty, that* the U.S. State Department invited each American professor to stay home instead; all of them acquiesced.

Imagine if Russia did not dissolve into cutthroat capitalism but adopted geonomics and showed the world how to thrive. I know, it looked scary to the 1%, too. By misguiding the new "democratic" government, the US elite dodged a bullet.

How is anyone supposed to hear about any of this? With all that money coming in, the current recipients of those rent streams cannot operate in total obscurity. Yet you must learn of their shenanigans in pages such as these.

A small number of corporations (themselves owned in leveraging amounts by billionaires) own the biggest broadcast networks, both radio and TV, and get to use the airwaves – which are public property– for free. If ABC News and CNN ever agree to merge, they'd reduce the number of independently owned national television news outlets from five to four. A new network has little chance of getting off the ground without agreements from Comcast and AOL Time Warner to carry it. The two companies serve about half of all cable households. With a handful of companies deciding all programming, many points of view are underrepresented. And they also own the largest circulation newspapers. They own the media which frame the mainstream worldview.

To paraphrase Lenny Bruce, "We, the media, don't have to tell you; we're a monopoly." Just as a small number of banks dominate mortgages. And a small number of owners in general own most of all valuable assets. And the same 1% dominate government and academia, also they master the media that keep land and rent invisible to the naive eyes of the 99%.

Lucky Gentry

Those present recipients of rents have gone to great lengths to remain rentiers. They fund the entire sphere of respectable research, keep politicians in their hip pocket, and buy up the media. By controlling the narrative, they consign unearned income due to our natural heritage to obscurity.

By controlling so much and keeping society in the dark, have wannabe masters of the universe (as some Wall Street speculators dubbed themselves) over-played their hand? Hardly. Despite needing to tighten their belts, the poor are not rioting. Despite needing to pop their pills, the middle class are not voting socialist.

One wonders why the wealthy bother. New knowledge is not brain candy for everybody. The old worldview suits most people perfectly satisfactorily, thank you very much. No news is good news, and good news is too good to be true. The gentry could just enjoy the irony – many people who could be inspired by the size of the worth of Earth are not curious to know the statistic.

Despite their strength and the public's blind spot, rentiers remain defensive. Even if the masses do not understand what's happening to them, the elite do. They know rent is a stream of spending that, as Bloomberg's Noah Smith notes, drives a way-wide wedge between the comfy folks and those struggling day-to-day.

Since favoritism plays the major role in amassing wealth, hard work and wise investment must play a minor role. Plus, the customary understanding of property allows profit from land to be a windfall for landlords, sellers, and lenders. Ergo, the very rich do not entirely earn their fortunes; they benefit from law and custom. More aware than the public of the source of concentrated wealth, the wealthy do still have to worry.

What might spell disaster for rent retainers is simple innate human curiosity, made itchy by hard times. People are aware of recessions, sensitive to losing their jobs, and a bit grouchy after being foreclosed from their homes. Eventually a few *must* put two and two together. That could increase demand for measuring our spending on land, resources, and other assets never created by anyone.

Despite academia being cowed, some economists must have felt curious. They became informed and published their findings, no matter how guarded their jargon may be. Let's analyze these articles in the academic journals. Then we'll peel off that layer and target the public agencies who should dish out answers for free and the private enterprises who could deliver stats that are precise. We'll uncover the rents that rentiers have covered up, and unearth (pardon the pun) the stories that they buried. Anything worth hiding by the few must also be worth knowing by the many, eh?

CHAPTER 13

SCHOLARS GUESS: HOW MANY $ IN THIS MARKET?

"The three most important things in real estate are not in mainstream economics."

STRAINING AT THE LEASH

To hear experts tell it, humans can do economies – produce wealth – out of thin air and without a place to stand. Or so conventional economists suggest. They have evicted the land from their discipline.

In a way, it stands to reason; central authorities do issue the symbol for wealth – money – out of thin air. However, they can only do so on their computer in the Federal Reserve, and that computer and that building must stand on land somewhere. Further, that location must be in the center of the nexus of power. In the US, that's split between New York and Washington, the locations of the Fed's HQ and its most powerful branch. If central bankers were not physically close to Wall Street and to Congress, they could not wield the power they do, which, as always, depends on relationships: being on a first-name basis with your fellow movers-and-shakers.

At any rate, given their attitude toward land as an antiquated, peripheral factor, many specialists today do not realize that taking an educated guess about the worth of Earth in America has an intellectual lineage. In years gone by, even mainstream academics did make an effort to calculate the total value of land and resources. Now, however, the institutional memory is lacking, hence this quest to measure all rents comes from more adventurous current researchers (a few such economists do exist).

THE ROAD TO ACADEMIC ANSWERS

We searched with the same key words as before: "worth of Earth" and "land value in the US" and "resource value in the US" and "rental value of US land," etc. But this time we entered them in the specialized databases for academic searches. The authors of those articles, at least, are respected by their peers and lay readers of economics.

All statisticians and economists in the field know how indifferent, even critical, most of academia and government are toward measuring the torrential flow of rents. For specialists to feel hesitant about hopping on a bandy bandwagon is a normal human reaction. Yet a few stalwarts did: and we salute them.

Even though those authors do address themselves to the rents paid for natural assets, they cannot do so in any other way than their usual academic fashion. Since specialists don't cater to us or offer any de-coder rings, we must meet them on their turf – the language of the specialists. So be it – the price of admission.

Academics and their dry style make us laypeople work. They focused mostly on land beneath homes and devoted pages of formula to torture the raw data into totals that might reflect reality. They crunched those numbers in their various "ivory towers."

Questions about the worth of Earth in America are a lot to ask of busy specialists who tend to have already said all they can say on the topic. Usually they don't have to deal with laypeople but only with other pinpoint-focused specialists fluent in the jargon. When a lay person seeks something as esoteric as the money society spends for the nature it uses, a specialist comes in mighty handy.

Academics have the stature that lets them confront their colleagues. They can:

- question the results of others,

- ask penetrating questions, and

- point future research in the right direction of where the total for rents lies. Some of them do come through.

While doing the research, it's a relief to see a help link at a voluminous website, or see a help desk in a specialized library. It's encouraging when bureaucrats point visitors to where the data concerning land values are kept in the buildings, in the stacks, and in the Internet. So their help is greatly appreciated.

As all authors say, books are a team effort. The help from specialists that has been forthcoming lets us advance this frontier of knowledge. And discover what dictums from special interests have been holding back the profession of economics.

We learned the definitions of their jargon and interpreted their columns, charts, graphs, and tables. Eventually, all that became clear. I waded through it so you don't have to. You only have to decipher mine.

Given that there are hundreds of thousands of articles out there in journal-world, the number that turned up on social surplus is not overwhelming. That says something about either the unattractiveness of land economics to economists or the kryptonite-effect from outside influences. Whichever …

What follows is the economists' best work, collated and annotated (to use their very own jargon). Given that their math is more obtuse than their text, let's spare ourselves most of the former and focus on the latter.

To find a recent total for the worth of Earth in America, I zeroed in on articles published in this millennium. Further, 2000 to 2019 is about the length of one period in the 18-year land-price cycle. However, I did make one exception for a forerunner who kept the torch of inquiry alive:

WHO TALLIES THE MONEY WE SPEND FOR THE NATURE WE USE?

"How Much Revenue Would a Full Land Value Tax Yield? In the United States in 1981, Census and Federal Reserve Data Indicate It Would Nearly Equal All Taxes" by Steven Cord in *The American Journal of Economics and Sociology*, July 1985.

As an academic, Cord estimated that 28% of national income in 1981 was land rent.

JS (author): As an activist, Dr. Cord was the long-time leader of the Henry George Foundation. He single-handedly persuaded dozens of towns in Pennsylvania to recover some of the socially-generated value of their locations. BTW, Pennsylvania is the only state whose constitution allows a legal class of cities to levy different rates on land and buildings. In a few Rust Belt states, a city, not a county, can do so but only with permission from the state legislature. In many other states, such as the growing states on the West Coast, speculators influenced legislators to amend their state constitutions to specifically prohibit splitting the property tax. They made the public recovery of socially-generated site rent, which stimulates infill, illegal.

Now we'll fast forward one score to this millennium. The land most people know – both specialists and lay people – is their yard. Plus, land for housing totals much more than land put to any other use.

65

TOP 10 STUDIES

The experts estimate the value of land in America. Their mileage may, and does wildly, vary. Their guesses follow in chronological order.

1. "http://www.foldvary.net/works/summary.pdf" by Dr. Fred Foldvary, Santa Clara University, Civil Society Institute (his website), January 2006.

Without giving the figures, Fred figures land rents could fulfill 60% of all public budgets. In 2006, the Census Bureau says that those budgets totaled $4.4207 trillion – $2.6554t federal and $1.7653t state and local expenditures; 60% comes to $2.6524t (almost identical to federal spending then). That amount would suffice if governments were to quit making income transfer payments. The beneficial results of treating land rent as the tax base, rather than taxing salaries, sales, and structures, include: *"higher incomes, reducing the demand for government welfare programs. Decentralization, privatization, and the elimination of wasteful government programs would further reduce the amount needed to fund government."*

JS: Well, of course, hats off to that, zeroing out waste. Yet if I may, instead of zeroing out food stamps, etc, one might prefer that politicians quit paying corporate welfare, fulfilling the wet dreams of weapons contractors, etc.

2. "The Value of Land in the United States: 1975 to 2005" by Karl E. Case (half of the successful and widely cited Case-Shiller Report), March 2007.

Ignoring agriculture, he estimates the total price of land in 2000 to be $5.6 trillion.

JS: The end of the last millennium (year 2000) was six years before the zenith and Foldvary's year of publication (#1 above) and 10 years before the crumbling of land-prices. Although his year is six years before Fred, Karl's amount is $3 trillion greater. One or both must be way off.

3. "The Price and Quantity of Residential Land in the United States" by Morris A. Davis (U Wisconsin-Madison Department of Real Estate and Urban Land Economics) and Jonathan Heathcote (Georgetown U, Federal Reserve Board, and CEPR), Nov 2006.

By the second quarter of 2006, the peak year in the recently concluded land-price cycle, residential land was priced at $11.6 trillion, more than double Case's year 2000 figure above ($5.6t), accounting for 46% of the price of the housing stock and 88% of GDP.

JS: But why GDP, which is annual, while land sales are infrequent and long-lasting?Is it just convenience? If using price, then comparing to other physical assets—maybe precious metals or museum pieces—might be more telling. Also, in 2006 home sites peaked, so while $11.6 t might look like a lot it might really be a little.

4. "The Price and Quantity of Land by Legal Form of Organization in the United States" also by Morris A. Davis (page 66), December 2008.

According to the Federal Reserve's Flow of Funds Report, the price of owner-occupied housing for the entire US in 2006 (the peak year) was $22.8 trillion – more than the capitalized value of the NYSE, Amex, and Nasdaq exchanges combined. The price of just the land underneath owned by households and nonprofit organizations was $6.9 trillion. The value of land owned by all four sectors – household and nonprofit, non-corporate, non-financial corporate, and financial corporate – at year-end 2007 was $12.4 trillion. The method that Davis and the Fed use is to take the price of real estate ($23t), subtract the replacement cost of structures ($16 t), to get the nigh $7t. Their methodology shows that land in the corporate sector [supposedly] lost almost all of its value between 1989 and 1995.

JS: While comparing price to price is the better comparison, Davis recalculated the value of residential land for 2006 from $11.6t (in #3 above) to $6.9t. For a "correction," that is enormous. They both can't be right and perhaps neither is. Further, how can land, especially downtown land where corporations have skyscrapers and locations are the most pricey, lose nearly all its value? One must wonder about their methodology.

5. "Commercial and Residential Land Prices Across the United States" by Joseph B. Nichols, Stephen D. Oliner, and Michael R. Mulhall, also of the Federal Reserve Board, February 2010.

According to the Federal Reserve Board's Flow of Funds (FOF) accounts, the total price of land held by households, nonprofit organizations, and businesses other than farms and financial corporations at the end of 2009:Q3 – close to the bottom of the recent land-price cycle – was roughly $4.5 trillion.

JS: Despite being close to the bottom, $4.5t might still be too low, since their high might not have been high enough. That aside, the drop of over $7t is huge. It shows that location values swirl while building values gradually depreciate.

6. "Over-accumulation, Public Debt, and the Importance of Land" by Stefan Homburg in *German Economic Review*, 15 (4)

In 2006, the US land-output ratio skyrocketed to $19.6 trillion in absolute terms, and then plunged to $8.8 trillion within four years, extinguishing over $10 trillion of actual or potential bank collateral.

JS: His 2006 peak of $19.6 trillion exceeds Davis's figure of $11.6t (#3 above), yet Homburg includes land used for more purposes than just housing while Davis looks only at home sites. Homburg's bottom of $8.4t exceeds the FOF nadir of $4.5t (#5 above) by nearly twice as much, yet the FOF figure includes more than home sites, so it's hard to see why the discrepancy. Further, Homberg's plunge extinguished well over half of his peak figure, while one half is usually the outlier amount of loss, claim many observers. So his figures bear watching.

7. "The Boom and Bust of U.S. Housing Prices from Various Geographic Perspectives" by Jeffrey P. Cohen, Cletus C. Coughlin, and David A. Lopez in the Federal Reserve Bank of St. Louis *Review*, September/October 2012.

For 2012:Q1, the aggregate price of all US land used only for housing was $4.151 trillion according to the S&P/Case-Shiller Index or $5.474t according to the FHFA Index.

JS: Both figures are well below Homberg's ($8.4 t) but in line with the FOF's ($4.5t). However, both figures are for 2012, two years after Homberg and three after the FOF. Given recovery had begun, it's curious both the FHFA and Case-Shiller lag so much.

8. "Aggregate U.S. Land Prices" by the Lincoln Institute of Land Policy.

They make two estimates of the total market price of land in residential use in the US for 2016 Q1, basing one on the public's FHFA at $8.746 trillion and and the other on private Case-Shiller at $9.940 trillion.

JS: Both estimates are less than the $11.6t for home sites in 2006, the peak year (#3), yet by 2016 housing had recovered its previous peak price. Once again, conventional statisticians lag behind the market. They're back to assigning to location only a third of total housing + site price while in 2006 they gave sites half of total property price.

Almost all the above articles focused on the value of land below homes. Finally in 2015 an official statistician broadened his target to include the values of land for all uses and other natural resources besides the surface. He also included all land whether it was owned by private parties or public agencies.

> 9. "New Estimates of Value of Land of the United States" by William Larson of the Bureau of Economic Analysis (in the US Commerce Dept), April 3, 2015.
> The contiguous (lower 48) United States plus the District of Columbia (1.89 billion acres), in 2009 prices could cost at least $23 trillion. The financial sector, or banking, which owned the least amount of actual land, had greatest value per acre, by far: 249 acres priced at $7.3+ million per acre.

> JS: It seems our work has been done for us. However, while it'd be nice to think that this is the ultimate answer, it's not. It's for 2009 – the bottom of the recent cycle – not for any year after, after land climbed out of the trough. $23t is not enough.

That article – and others like it – ends with a plea to colleagues to do follow-up research. If any academic did further research to reach an exact amount for rents, no journal I checked published it, regardless how much we need a sound measure.

None of the above top nine articles crossed the finish line and tallied the holy grail of the worth of Earth in America. Either they left out public land or left out land used for something other than houses, or based their guesses on combinations of land plus buildings or used out-dated stats. If the rentier powers-that-be tried to discourage inquirers, they succeeded.

Yet whatever official estimates lacked in accuracy they made up in respectability. Almost everyone who finds cause to cite the value of land cites the official figure, warts and all, as gospel, without bothering to question the method used. So let us get our own hands dirty and question that method – by taking a look at rich cities.

The low figure for land—and high for buildings—follows from the method the mainstream uses to separate the value of land and building in a combined price for a property. The value that conventional economists give the value of buildings of any age is their replacement cost. The value they give to land or location is what's leftover.

A couple of mainstream economists offered a correction. Nicolai V. Kuminoff and Jaren C. Pope write, "First, the replacement cost approach may overstate the value of land during a boom-bust cycle. Second, the bias may not be neutral. Our results suggest it would be largest in the highest-amenity neighborhoods." (This is from "The Value of Residential Land and Structures during the Great Housing Boom and Bust.")

Pope and Kuminoff also "suggest that moving from a property tax to a land tax may actually help to stabilize revenue streams for some municipalities." Shades of Steve Cord (see above)! Does even such faint praise unnerve rentiers so that they frown on research into the value of land or, if one must, at least distort it? Consider again the conventional ratio of land to building. Specialists have a hard time accepting the fact that in the wealthy ski resort at Aspen CO a vacant lot fetches $10 million easy. A million dollar house in San Francisco is actually an $800,000 site with a bungalow on it. Same goes for $10 million dollar apartments in New York; those are $8 million sites. Developers pay $1,000 per square foot and upwards in any world-class city like New York, London, or Tokyo. That square foot is for the floor, drawing attention to the building, but includes the site, the part overlooked. Remove the building and maybe it's easier to see.

Fortunately in our quest to find a realistic total, a few economists did grapple with the twinned issues of doing the research while going against the grain – and did so by also looking at cities, both rich and poor, both big and small.

> 10. "Metropolitan Land Values" by David Albouy, Gabriel Ehrlich, and Minchul Shin, 2017 June 11.
>
> The Federal Reserve in its Flow of Funds (FOF) accounts uses a method to determine the value of land that yields *negative* values, which cannot be right. Instead of switching to a more accurate method, in 1995 the Fed stopped publishing its estimate of actual land value, yet continued to publish a total for land and buildings combined.
>
> Rather than try to derive a value for land from that stat, the Albouy team turned to records of the sales of land alone. They tallied 68,756 land sales covering 76,581 square miles of urban land, or city regions, that public statisticians named "Metropolitan Statistical Areas" (MSA). In 2000, all MSAs accounted for 80% of the US population and probably at least as much of the US land value.
>
> Most Americans choose to live in or near cities, with more coming. Cities have magnitudes higher location values than the countryside. Rural areas have most of the land while urban areas have most of the land *value*.
>
> High ratios favoring land give conventional brains vertigo. Mainstream statisticians in 2006 (the peak year) tabulated all real

estate at $43.3 trillion, structures at a whopping $26.3t, and land at only $16.9t. Albouy and his team could deal with the vertigo. Using land sales, they calculated its aggregate price was $30.4t, nearly 80% higher.

Also, their estimates proved more stable than those of the FOF. At (or near) the bottom of the last cycle in 2009, the land price total fell, according to the FOF, down to only $5.8t yet, according to Albouy *et al*, down to only $14.4t. The peak-to-trough decline in the FOF was 66% – well beyond the normal outlier of 50% – while in the Albouy study it was 40% – well within the 50% normal outlier.

How do they reconcile the huge difference between official results and their own? First, as cities get bigger, their sites grow in value, more so than do their buildings. Officials overlook this phenomenon and subtract an exaggerated value for buildings. Second, the FOF, while including land outside metro areas, left out public land. The Albouy team, while excluding land outside metro areas, included land for civic buildings, parks, and roads. Assuming that the public owns urban land worth 40% of Albouy's total, then private parties own land priced at only $18.2t of the total, which is much closer to the FOF $16.9t.

JS: Finally, the 10th of the Top Ten provides a figure that's ample and accurate. Albouy's $30.4 trillion for only metro land vastly exceeds Larson's $23t for all other kinds of land. And the larger number is derived directly from actual selling prices of land.

The other nine academics slanted land value downward. The authors bent over backward to minimize the object of their study (the value of land). The economics discipline, literally, constrained its mildly wayward members. It was as if the authors served not objective science but some interested party. To my ears, the tone of the articles did not have a ring of scientific caution in the face of the unknown but of political caution in the face of known biases – those of the ruling rentiers.

The best figure for us for 2006 is Albouy's $30.4 trillion. Since real estate recovered all its lost value by 2016, in 2018 the total aggregate price for all land was well over $30.4t. Furthermore, there's still the rural land to add in, which we'll tackle in Chapter 15.

These eleven studies are what individual academics had to say. Next let's dig deeper and visit the economists's official sources – academic centers and government agencies – for any missing rents, especially for natural resources. De-coder rings ready?

CHAPTER 14

MINING NUMBERS AT OFFICIAL STOCKPILES

Officeholder: "What's the correct answer?
Economist: "What do you want it to be?"

PEERING INTO THE EXPERTS' SOURCES

Backtracking is the name of the game. We began with popular articles by journalists who hyped a total for the worth of Earth in America (in normal English), citing the findings of a researcher (Ch 11). Peeling the onion, next we analyzed the academic articles (in professional jargon), whose authors based their findings on official stats (Ch 13). Now we'll pay a visit to the stocked ponds where specialists go fishing.

Our authors used the academic centers and government agencies that record prices for real estate, sometimes even for land alone, or sometimes even for natural resources. Scouring those official sites and more, we draw closer to figuring out the total value of assets never produced by anyone's labor or capital. Knowing the total's change through time, we would know the business cycle better. And we realize a flow of dough that's a surplus, socially generated.

Another reason we need to plumb the wellsprings is to keep current; the flow is always flowing so the total never stays the same. Whatever amount was reported before won't be accurate now. To know that number requires us to harvest the latest available stats.

Plus, we find out if the statisticians at bureaucracies and enterprises have become more or less accommodating since investigators last visited them.

My list of entities to query about natural assets has about 80 names on it. Given that phone numbers change and the person who answers typically transfers the caller to someone who may know the answer or transfer you yet again, that'd total well over 100 calls. I decided to rely on the new custom of email.

I asked all the departments of real estate in all the universities in the US (that I could find) for a figure on "the total value of land and resources" as they did not understand "Earth's worth." Some institutes are in the Ivy League, some are in business schools. I contacted all five dozen of them – and dredged up some new facts.

Some think tanks that think about ground rent are around Boston (America's college central): the National Bureau of Economic Research and the Lincoln Institute. Others are in the nation's capital, Washington DC – the Urban Land Institute, the Brookings Institute, the Tax Foundation, and the Center for Economic and Policy Research.

- NBER, the granddaddy of them all, has been making scholarly attempts at totaling the worth of American land going back over a half century, but nothing current.

- The Lincoln Institute updates their database every quarter with the latest (albeit still lagging) output from three deeper sources, one private, two public, but nothing ancient.

- The Tax Foundation had lots of links to deeper sources, but no summaries.

- Center for Economic and Policy Research (CEPR), the young upstart, greets one with a site that looked the best, but looks are not everything; here, looks are the only thing if you're looking for rents.

While a goodly number of the 80 replied, none gave a satisfactory answer. Whether public or private, there is no service, institute, office, department, bureau, board, or administration that's assessing all land and resources or tracking all rents. Out of all of this nation's enormous bureaucratic and academic infrastructure, nobody is curious enough to record how much society spends for the nature it uses.

It does not come as a surprise, but more like a shock. How can the professionals be so indifferent, suffer such a blind spot? What's so off-putting about this line of inquiry?

At least a few agencies did focus on what people put on top of the land – buildings. Those stats serve as a proxy that we can extrapolate from.

Top 10 Agencies: Looking for Rent in Mostly Wrong Places

1. Housing and Urban Development: Whereas the other federal agencies focus on houses (where the middle class live), HUD focuses on apartment buildings (where, mostly, the poor live). HUD pays many of those poor a voucher good for housing. In 2012

(HUD was six years behind – imagine a bureaucracy doing that), HUD counted 2 million properties, yet the National Apartment Association says there were 2.3 million back then (if you can't correct them, still cite them). HUD found rental receipts to average over $100k per building and to total $200b for all buildings for the year.

JS: HUD numbers come from apartment owners filling in a survey, and the less value they admit to the less tax they're required to pay, so the actual rental value could be higher. HUD gives the building rent, an annual figure, and we're seeking an annual figure, albeit for land not buildings. The average selling price of a complex was a bit over $1 million. With receipts averaging a tenth of that, we derive a ratio of price to rent ratio of 10 to 1. So the value of the land underneath apartment buildings would be at least $100b in 2012. The owners' property tax was under 1.5% or $3b. The land half of that, or $1.5b, gets added to site rent, too.

2. Federal Housing Administration: Part of HUD.

JS: If the FHA has any unique data, they bury it somewhere.

3. Federal Housing Finance Agency: The FHFA provides no totals but has a House Price Index covering from 1975 to 2017. They say in 2017 housing cost 5.7748 times what it did in the mid 70s.

JS: Use Case's figure for 1975 land value – $291,740,000,000 (Ch 13) – and multiply it by the FHFA figure. Then land value in 2017 comes to $1,684,740,152,000. A tad more than a trillion and a half is quite small compared to most estimates that went before (Ch 13). Which figure is likely off? Case's land value or the FHFA's multiplier? In 1975, the stock market was $137,281,000,000; if Case's number were any bigger, it would be outlandishly bigger. That leaves the FHFA multiplier as the culprit, which makes sense. In some of the towns where I've lived the last few decades, housing does not cost merely five or six times as much but 10 or 20 times as much.

4. Freddie Mac: Like the FHFA, they too have great looking tables but no totals.

JS: At least they have an easy-to-use "ask us" form; however, the answer was not overly enlightening.

5. Fannie Mae: No tables, no current data, no easy contact.

JS: But they were kind enough to ask me to call. They told me that they don't even have the total for the land (and buildings) that they own! (Your tax dollars at nap-time.)

AGENCIES REPORTING HOUSING+

The US Department of Commerce has three relevant bureaus:

6. Bureau of Economic Analysis: The BEA gives four relevant measures: personal spending, rental income, GDP, and fixed assets. At the end of 2017, for personal spending, the BEA listed housing but combined it with utilities, coming to $0.120,442t. They gave rental income $0.757.4 trillion. At the end of 2016, in their GDP by industry they gave real estate $3,454.6t; agriculture, $0.4281 t; and mining $0.3856t; totaling $4.2683 trillion. They gave fixed assets $20,785.7 trillion.

JS: Of course, these figures combine the values of both buildings and land.

• Plus, in the BEA's personal expenditures, now it's not just land complicated by buildings but also by utilities. These bureaucracies are heading in the wrong direction, away from a stat for land alone. The BEA's figure for housing in personal spending is way below the 30% or 40% or 50% of income that most studies cite and differs wildly from that of Labor (coming up below).

• As for the minuscule figure for rental income, recall that it's not from actual figures but from landlord surveys.

• Their GDP figure, if halved, is well below our previous figure for a flow, Foldvary's $2.6524t and that was 2006, ten years earlier (Ch 13).

• And the BEA's number for fixed assets, despite adding whatever's on the land, is way below Albouy's $30 trillion for land alone.

Lots of inexplicable contradictions.

7. Bureau of Labor Statistics: The BLS uses a "consumer unit." In 2016, the US consisted of 129,549,000 of them. Their average income was $74,664. On average that year they spent: on food, $7,203; on housing, $18,866; on utilities, $3,884, on vehicle fuel, $1,909.

JS: Those categories, food to fuel, are the ones with the biggest portions of land or natural resources. Of course, the price of everything includes a portion for how much the seller or producer paid for a place to do business and perhaps raw material(s).

To compare to the BEA (#6 above), housing and utilities came to $22,750. Times the number of units, it comes to $2,947,239,750,000 – well above the BEA and well in line with most studies on how much income housing consumes.

These four BLS categories in 2016 totaled $31,862. So, consumer units spent 42% of their income on these categories. If half of the value of those categories is rent, then they indirectly spend 21% on land, locations, and resources. As a whole, the 129,549,000 units spent $4,127,690,238 on those four rent-stuffed categories. If half that figure is for land and natural resources, it still seems low. Of course, all this leaves out spending by nonprofits, business, and government.

8. Census Bureau: For 2016, they gave an aggregate price for all-housing units: $21,935,096,166,400. For "real estate andrental and leasing" they tabulated a revenue of $487,655,249,000. For the total revenue that states raised via taxing property – land and buildings – they tabulated $19,031,950,000; 14 states were left blank and some of those states were big and rich, like New York.

JS: Their $22 trillion for housing was close to the BEA's $21t for fixed assets, but you'd think that all assets would be greater than just one asset. Whatever, the Bureau's return – under a half trillion = 2% – is abysmal. Obviously, their return leaves out "imputed rent" or the value of owner occupancy. As for the property tax, the revenue collected by states was even less than 1% assessed value (price)—pretty paltry considering society generates the value.

The above are the eight federal agencies that deal with residential costs. There is a ninth organization that is not exactly a part of the government nor is it apart from the government. It's in a limbo land – like NBER, which receives federal money and supplies government with many of its bureaucrats, both low- and high- level. I refer to the self-christened ...

9. Federal Reserve: It's a private corporation but at the same time chartered by Congress to, ostensibly, control inflation and unemployment (Chs 12 & 26). They release their Financial Accounts of the US. It has Flows and a Balance Sheet, most recently for 2017 Q4.

In Flows, they created a category they call "Households and nonprofit organizations; gross fixed investment, residential equipment and structures (includes farm houses)" and another "Nonprofit organizations; gross fixed investment, nonresidential structures, equipment, and intellectual property products." Let's assume their structures sit on land whose value is folded in. The Fed priced structures at $632,900,000,000 and land at $148,700,000,000. Together that's $781,600,000,000.

In Balance Sheet, they created "Households and nonprofit organizations; real estate at market value" they priced at

$27,848,300,000,000.

JS: Their $28 trillion is more credible, greater than all previous official figures. It'd be higher with the real estate of corporations and governments, yet lower with land alone (excluding improvements). Their Flow from that asset (about $800 billion) is also only 2%. The mighty Fed leaves a lot unanswered.

In sum, those nine agencies, with different sources and different definitions, create tables nearly impenetrable – the old priesthood syndrome – and largely irrelevant, unless minutiae are your thing. They yield conflicting lump sums and make no effort to explain why they vary from their brethren; did they all use rubber rulers? Other fields have agreed on definitions – like the different kinds of clouds in meteorology, so why can't our public statistician and economist servants be as considerate?

Those official number-crunchers might live off public money but they write and format for each other, like school kids sending notes in code to each other during class. How can they take our money for such a performance? Will anybody raise the bar? The public needs to know.

UNOFFICIAL ANSWERS – MORE ACCURATE?

Officials and academics sometimes buy data. Let's see if the private companies deliver any better results. They might have more incentive to nail down the data exactly. Investors like to know true yields. And, unlike officials, companies have to be user-friendly. Or they go broke.

I could not find the answer at the National Association of Realtors, the long-standing go-to group. Nor at CoStar, which a lot of researchers use, but they charge for services. However, I did get a stat at the site of the new kid on the block. Their website – out of all the university, government, and realtor sites – was by far the prettiest and easiest to use.

10. Zillow: "Total Value of All U.S. Homes: $31.8 Trillion. How Big Is That?" By Zillow Research on Dec. 28, 2017.
    Renters spent a record $485.6 billion in 2017, an increase of $4.9 billion from 2016.

JS: How much renters spent is pretty close to the Census Bureau real estate revenue (#8 above). However, this privately calculated figure for home prices beats all the totals that public agencies tabulated. And it's just for homes, not for any other buildings sitting on land.
    Zillow used to show the separate values of home and land in the Tax Assessment area in the record for each property. *They no*

longer do. Now the only value they show is the combined value of location and improvement. Land value is no longer available from Zillow because even they cannot get it from any agency, from local assessors to federal Census Bureau.

If land is half of Zillow's near on $32 t, then residential sites might approach $16t. If you substitute that $16t for the figure Larson used for home sites in 2009, then Larson's $23t (Ch 13) total goes up to nearly $30t for all land. That's in the ballpark with Albouy's figure of over $30 trillion of a decade ago for metro land (Ch 13). Thus Zillow does lend credence to Albouy's total and methodology.

Most researchers above leave out lots of land. Humans don't just put houses on land. We erect other buildings, too – offices for commerce and factories for industry. And we don't just build, we also pave, yet most tabulators overlook that land used for streets and parking. And some land we don't build anything on but take something out, as we do from farmland, forests, mines, oil wells, etc. And once again, we won't leave out a patch because it's owned by a nonprofit or business or public agency.

To tally a true total, the category "Land" would have to include both urban and rural, solid and fluid, and private and public. We found some figures for those categories, however accurate or complete they may be. We'll play with the cards we're dealt, work them into our calculations, and get closer to that grand total for the worth of Earth in America.

CHAPTER 15

LAND: THE MORE PRISTINE, THE LESS PRICEY

Oil makes some rich, topsoil not so much, but every trillion counts.

"ECONOMIC LAND" CAN BE WET, BREEZY, EVEN ETHEREAL

Like weekend vacationers, from urban to rural we go. To complete the tallying of the worth of Earth in America, let's rope in the remaining rents found outside the city. In the countryside, humans utilize land to farm, to graze, to log, to mine, to drill, etc. On the cadastral map for an entire region, you see the labels for all the uses: residential, commercial, industrial, agricultural, pastural, sylvan, mineral, etc.

Besides water, there are other fluids, like oil, plus the electromagnetic spectrum. Bigger picture, we must not only count the surface (like home sites) and subsurface (like oil fields). We must also count the supra-surface – the airwaves and soon the geosynchronous orbits.

Furthermore, land is not just acreage owned by individuals or households or families but also by nonprofits and for-profit businesses and corporations, plus governments. Most of the lands not yet counted are owned by the public, via our governments. You'd think we the people might like to know the value of our natural holdings. And that our public servants would be able to tell us.

TOP 10 PUBLIC TALLIERS OF PUBLIC LANDS

1. The USDA's Economic Research Service, unlike the USDA's National Agricultural Statistics Service, stepped up. They projected farm real estate to be $2.6 trillion for 2020. Minus buildings, etc, worth at most 20%, crop land and pasturage together come to over $2 trill. BTW1, owners rent out 40% of farmland; non-owners worked over half of cropland and over a quarter of pastureland (reminiscent of feudalism). BTW2, over the last 20 years, we've lost 31 million acres of farmland – about the size of Iowa – three acres every minute (human diners, take note).

JS: It's not just farmland whose value is judged by annual output, i.e., the harvest. Commercial sites, too, are determined by how much money merchants can make at the location each year. As they do for metro regions, some sources – like NASS above – give a figure for land and buildings, not land alone. For those estimates, researcher Merijn Knibbe suggests "applying a 20% haircut to distract the value of farm buildings" (in real-world economics review, issue no. 69) to calculate the value of farmland alone.

So 80% of NASS's 2016 figure is $2,115,384,800,000 trillion. For 2017 using ERS's figures, $3k times 911 came to $2.80588 trillion. That's a pretty big one-year jump but the bigger number is more recent and used smaller aggregates which tend to be more accurate. Adding it to metro land (Albouy, Ch 13) puts the total for both city and country at $33 trillion.

Moving from the big gorilla of private land to the little monkey of public land, value-wise. Acreage-wise, the US owns about 1/3 of America and from it raises revenue. Much public land is "unimproved." Without much in the way of improvements, not subtracting their near-zero value does little to alter the combined total. In the boonies, almost all the value is from the land and resources. Life becomes easy, one hopes ...

2. The Bureau of Land Management, in the Department of Interior, has its National Integrated Land System (NILS) GeoCommunicator.

JS: A great name but I'll be dogged if I can find any land values there.

BLM also has "Public Land Statistics" which shows at Table 3-25 that the BLM, from its 1/3 of America, in 2014 collected for us citizens $0.344t in 2014 and in 2015 got for us only $74 billion.

JS: Billions? That's not much, and a huge drop from 2014 to 2015. Gee, how'd you like to have them as your steward; well, my friend, you do.

3. The Treasury has a Bureau of the Fiscal Service which produces a Financial Report of the United States Government. On Balance Sheets under Assets it has "Property, plant and equipment, net." For 2017, they put it at $1,034.5 billion (p 10 or 17). Later they separate out land at $13.4 billon (p 83), exactly same

amount they gave to the cost (?) of land. "Property" probably includes "Federal Oil and Gas Resources" and "Coal," but if not they found only $47 billion for the former (p 197) and $9 billion for the latter (p 199).

JS: In their scheme, land is a tiny fraction of property. One reason is they impute a huge cost to land. Of course, there is no cost to create land, so it must be what they're doing to land. Further, their $47b is well below the BLM's $74b. Both are too minuscule to be realistic. Neither would add much to the growing total, already in the trillions.

4. In the basement of the White House is the Office of Management and Budget. The OMB produces its "Analytical Perspectives" that has a section called "Federal Receipts."

JS: While we could not find revenue from land or resources, a scholar told us he could for an earlier year; his estimate then was about $0.5 trillion.

5. The General Services Administration (GSA) created a database they call The Federal Real Property Profile. FRPP presents its "FY 2016 Open Data Set." In there is "Table 1: US and US Territories." For FY 2016, they give land, excluding public domain land (which is huge), a cost, but not a value.

JS: How can an official "profile" or "data set" be so incomplete?

Not to be outdone by the executive branch, the US Congress also has its agencies who could have stumbled across the value of the land that makes up America.

Congressional Attempts – More Data Vacuum

6. The Government Accountability Office (GAO) notes whenever a part of the government behaves irresponsibly, such as officials keeping lousy records. "About 34 percent of the federal government's reported total assets as of September 30, 2016, and approximately 18 percent of the federal government's reported net cost for fiscal year 2016 relate to significant federal entities that, as of the date of GAO's audit report, were unable to issue audited financial statements…"

JS: An earlier GAO report gave a total for federal land but no key word is turning it up now.

" ... significant federal entities ... were unable to issue audited financial statements ... "
— The Government Accountability Office's website

7. The Congressional Budget Office in the past estimated that urban dwellers spend over 7% of their income on residential land and at least 4% of their income on other land costs.

JS: That'd be about two and quarter trillion dollars. Of course, country folk spend some of their income on home sites, too. And it's not just housing costs – we pay for land whenever we pay for anything. When you pay for food, part of that payment goes for farmland. For gasoline, part of that payment goes to whoever owns the ground with oil under it. ... And so on.

8. The Congressional Research Service has published such reports as "Federal Land Ownership: Overview and Data."

JS: Despite such an intriguing title, nowhere inside did it give the total value or price of all federally owned land.

9. The Library of Congress, of course, contains just about every word ever published.

JS: Ever try to wade through anything like that? The proverbial needle in a haystack. But the stack will still yield something of value eventually, one hopes.

There are more public lands whose location values may not have been included. The list includes: port districts, landing slots at airports, ship berths in harbors, boat slips in marinas, highway, bridge, and tunnel tolls, etc. While it's a pretty extensive list, in the bigger scheme of things the values are not so huge that if they were overlooked, it'd not undercut the grand total that much.

Besides *terra firma*, there's water; the 2nd most essential element of life. Every moment, we need air. Every day, every living thing gets thirsty. One hopes the given values of land and resources above would include access to water (where applicable), but they may not. If statisticians left water out, here are some figures from elsewhere:

10. The US Geologic Survey figures Americans in 2010 withdrew about 355,000 million gallons per day (Mgal/d) from surface lakes, rivers, etc, and from the underground water table. BTW, the biggest user of water – bigger by far than second place agri-business, bigger than farms, fac-

tories, mines, and homes put together – is *power plants generating electricity* and thereby needing cooling (which may be another good reason to go solar). Yet the USGS did not give a price. The University of North Carolina's Environmental Finance Center did: *one cent per gallon.*

JS: Multiplying by 365 days in a year, water prices out at $472,948,750,000. Given that both the population and economy has grown since 2010, and that one penny is the absolute rock-bottom figure, it's easy to see a realistic value of $0.5 trillion for water in 2018. Paid annually, that's not a price so much as a rent. Yet, that's for water out of the tap, while we want its long-term value, in the ground. For its value *in situ*, let's estimate a quarter-trillion bucks.

None of these results for the value of income from public land is really much help. But even these spotty returns *do* all add up.

Do Agencies Check Their Sources?

After all this – 10 official rural sources, 10 official urban sources, 10 academic articles, 10 popular press articles – what can we conclude?

For some public land, our public bureaucracies offer no totals. They leave it to the curious to add up the subtotals. Worse, the estimates by various agencies differ wildly. Which one of their competing figures is the most official and the most accurate?

Worse still, some of the official figures are suspiciously low. We the People (US/State/local Governments) own over a third of America. So, unlike one's private land, our public land can be so unprofitable?

In fact, no. The public agencies only count the rents and royalties they collect. They overlook the royalties *they fail to collect.* When one adds these unpaid still owed amounts, then the given value of public land rises appreciably.

A few episodes of large-scale corporations cheating the public out of their share have been reported by the mainstream press. Are those incidents the only ones? Or are they just the tip of the iceberg? How are we to know? And without knowing, how can we know the real worth of public Earth?

I am shocked… not really by the predictable greed and cheating of the unrestrained very rich. But, also, by the complicity of our public servants with a feeble enforcement system, compounded by their bad bookkeeping. They keep silent about the elite's breaches of contract, and silence is tacit consent. Then they declare that their total – missing so many data points (facts) – is the true, accurate total of the value of public land. To put it politely, doing so is at least sloppy and misleading, at worst seriously dishonest.

5 Best Private Tallies

All of the above except the US Geologic Survey get their info from assessors and appraisers who get their figures from actual sales (which is where the UNC EFC, above, get theirs for water). Since tabulating land value is their business, perhaps we can find a total at a professional organization.

11. The International Association of Assessing Officers has a great tag line: "Valuing the World." So if anybody should know the worth of earth, it must be them, right? Well, their librarian might tell a member assessing officer, but not just anybody, at least not riffraff who aren't paying any membership dues.

While assessors do their guesswork for government, appraisers do theirs for business. Let's question those whose pay depends on their performance for their clients.

12. The American Society of Farm Managers and Rural Appraisers is very newsy and releases farmland aggregate farmland prices, but only to members.

13. The American Society of Appraisers has a nonprofit mission and they claim to work from basic economic theory and legal precedent. Plus, they're the ones everybody else gets their numbers from. But they've not made available a solid number for the price tag of America's land if it were for sale. Or the rental value if it were for leasing.

While that appraiser/assessor well is dry and despite the shoddy performance of our public agencies, I'm strangely buoyed. While our public servants and public savants bow to pressure and crank out distorted facsimiles of reality, others didn't. Here are two.

14. Purdue's Kevin J. Mumford in "Measuring inclusive wealth at the state level in the United States" provides inputs that come to $533,305,319,000 in year 2000.

JS: His well over a half trillion dollars is much less than farmland in #2 above, likely due to the passage of time – 17 years. Over almost the last two decades, the acreage of rural land has decreased – due to sprawl, mainly – yet that very sprawl has driven up the value of the remaining rural acreage, especially close to metro regions. So, the value of non-farmed rural land would be significant but still unknown.

15. The Heartland Institute's Richard Ebeling wrote "There is No Social Security Santa Claus" (2015 Dec 22). He noted the Feder-

al Government owns mineral reserves of copper, nickel, gold, zinc, platinum, lead, and silver, plus 257 million acres of grazing land and 250 million acres of timberland. He puts the price tag for all those natural federal assets at $5.5 trillion. He said he used many official sources which, however, he did not name.

JS: Interestingly, $5.5t is also the new record for all central banks' bonds that no longer pay interest but charge it against bondholders, to put the amount in context. So if Ebeling's $5.5t is accurate (and he did do thorough research) that's a hunk of change. It's overwhelmingly immense yet overlooked and set aside by academics, bureaucrats, and journalists.

It would push the $33 trillion above (Albouy's plus farms, mainly) to $38.5t. The uncounted non-federal, non-agricultural rural land could push the total to $40 trillion. The passage of time would by now (2019) have pushed it to the the mid 40s, nearly double Larson's $23t (Ch 13), and at 10% yielding a rent total around $4.4 trillion.

Recall the GAO's gentle critique of official figures (#6 above). We're not the only ones to notice flaws in public bookkeeping. We'll see what others have said about that. And add a new word to our list of key words: critique. We, at least, are free to follow the facts to wherever they may lead. No torpedoes for us to damn; full steam ahead to the truth inconvenient for some but liberating for all.

CHAPTER 16

EVEN A WORLD BANKER
RIPS OTHER ECONOMISTS

"Don't tell my Mom I'm an economist. She thinks I'm a pianist in a brothel."

BOTH SIDES SHARE THE SAME BLIND SPOT

At least one conventional voice howls in the academic wilderness (we howl *at* the academic wilderness). Lars Peter Hansen, who in 2013 was granted the ersatz Nobel Prize (actually, the global bankers' prize), said, *"I believe that the recent financial crisis exposed gaps in our knowledge."* His colleagues' response was, itself, a gap. Mainstream economists seem nonchalant about their own colleagues critiquing them.

Recall Karl E. Case, one half of the go-to team for housing prices – the Case-Shiller Report – noted a specific gap in academic knowledge, the one with reference to land rent totals (Ch 10), never mind the specialists tabulating a good number for the worth of Earth in America. Both Albouy writing about metro land and Larson about all land (Ch 13) make the usual plea for collegial follow-up. Yet their monumental works go uncited by their peers. (We mention their figures to insiders every chance we get.)

Everywhere you look, land, land everywhere, but not a plot to count. Yet land's rent is big and powerful – everything you'd want in an economic phenomenon; it is many trillions and it drives the business cycle. But economists don't measure it and don't theorize with it. That's two strikes – one away from being an out.

While we find economists asleep at the wheel – most of them overlook land, one of the only three factors in production – actual members of the discipline find many more faults. According to those economists, their cohorts get nearly everything wrong, from basic assumptions to technical writing. Yet ironically, even the critics miss land, the discipline's most blinding blind spot, just as the "complacents" do.

Are the two phenomena – absent land and crippled economics – linked? Is the reason why economics frustrates even economists the fact that they lost land as a factor? So that now they founder and devote themselves to distractions? And code their claims in impenetrable jargon? Thereby facilitating the adoption of mistakes as gospel?

WHAT'D THEY SAY?

Insider critics of jargon can occupy lofty positions – at least for a while. Paul Romer, Chief Economist of the World Bank, criticized "Bankspeak," the jargon of his underlings, and typical of practitioners of the discipline in general. Romer's own boss, the head of the World Bank, responded forcefully and, instead of supporting more clarity in exposition, stripped away some of Romer's job duties.

In 2015, Stanford University's Literary Lab found the Bank's writing was "codified, self-referential, and detached from everyday language." They're the ones who coined "Bankspeak" for the lender's "technical code." They also noted that Bank authors would link long chains of nouns with the word "and," thereby producing mind-numbing lists.

Among his sins, Romer:

- imposed a quota on the conjunction "and,"

- canceled a regular publication by the World Bank that didn't have a clear purpose,

- insisted that presentations get straight to the point, cutting staff off if they talked too long, and

- let his subordinates know it'd be a good idea to dive right into public debates and align their work with the goal of the Bank – ending extreme poverty and reducing inequality.

The Development Economics Group (DEC), the research department of the lender, issues reports composed in a dense, convoluted style. Romer urged researchers to write more intelligibly, using the active voice to be more direct.

He erred in focusing on precision in communications and not on the feelings of economists. The more than 600 DEC employees pushed back and the Bank president sided with them. Now Romer no longer oversees the DEC. Nor, if you've been waiting for the other shoe to drop, does he work there at all, after speaking frankly in an interview.

The clipping of Romer's wings reminds economists that even prominent ones – they awarded Romer the so-called "Nobel" Prize – cannot rock the boat without consequences. If they can't even raise the bar for communication, how could they consider doing something as fundamental as researching a basic factor in production, i.e., land? It seems they can't.

Bad as it is, what World Bank economists crank out is as good as it gets. Members of the discipline rank the World Bank as tops in development research. In terms of the number of times its output is cited, they're first, ahead of the London School of Economics, Brown University, and Harvard University.

JARGON JUSTIFIABLE?

Another economic institution has dealt with the jargon problem not by handcuffing a critic but by revising its writing. That's the World Economic Forum's Global Agenda Council on Space Security. They're publishing a book to explain in plain language how space-based technologies and services can help society face its greatest challenges.

The Guardian's John Lanchester lists his favorite (least favorite?) jargon. One is "quantitative easing." It's a euphemism to cover up the fact that the biggest central banks create brand new money from nothing and hand it over to lesser banks and friends until they can pay it back. One main way the recipients avoid bankruptcy and amass funds to repay the gift is to buy government debt and ship the interest to the central banks. Able to count on banks to always buy their bonds, governments slide (or rush) deeper and deeper into debt.

A more colorful phrase was the chairman of the Fed's suggestion, citing libertarian Milton Friedman, to toss cash out of helicopters.

Outside the helicopter, we find professional presentations of their findings to be tough sledding, hardly packaged for we laypeople. Our being uninitiated non-specialists on the DIY kick, we strain the eyes peering at tables and reports and articles. We wade through all the numbers, jargon, and faulty reasoning, sparing others this initiation rite.

Sure, jargon can be used to convey thoughts to other specialists. Yet writing for peers means not writing for we lay readers. Given the caution with which economists must tread, being obtuse is a way of flying beneath the public's radar, and thus the elites'. Writing in jargon is better suited to cover-thy-butt than to revealing important truths.

Jargon serves as a code to exclude non-specialists. It creates insiders and outsiders, becoming a badge of belonging. Every generation and every discipline does it. Humans are wired that way. And cliques are fine for teens, but for those supposedly adding to society's storehouse of knowledge? Most academics are paid – directly or indirectly – from members of society paying taxes. If taxpaying lay people cannot understand their tax-paid academics, how's that fair?

More irony, besides insiders objecting to their discipline's jargon, is that pointedly ignoring land could be the root cause of corrupt economics. Without land rent to include in their calculus, economists cannot calculate logically – i.e., distinguish between spending for the produced and the non-produced. Without much of importance to say, they still say much, and say it guardedly. Academic economists can't shake themselves free of the habit of expressing themselves in ways to avoid impolitic mistakes.

CLARITY, A *SINE QUA NON*

Not only do they speak mainly to each other, but they have few outlets through which to speak to each other in. If they can't get published in one of less than a half dozen journals, then they cannot advance. And the journals get to control the debate and the topics debated. Well, if denied by their superiors, maybe they should be talking to we of the *hoi polloi* anyway. Figuring out how to make sense to the intelligent uninitiated could do economists good.

Life in the cloister insulates one from having to deal with challenges affecting a majority of humankind. To reach the goal of sensible specialized writing, *"we also need to connect our academics to the real world rather than trying to free them from it,"* suggests Jack Stilgoe, a Senior Lecturer in the Department of Science and Technology Studies at University College London.

Scientists going at each other, debating opposing points of view, is par for the course. It's one way theories and hypotheses get tested and perhaps proven; part of the scientific method. Fortunately for real scientists, they can slug it out in the laboratory or in the field. Unfortunately for economists, they can't. Unable to run experiments, they need to write clearly in order to think clearly.

Writing non-clearly allows and reinforces muddled thinking, which helps explain why economics in particular and social studies in general are such a mess. So, universities make scholars study statistics and other things that their phones can do for them. What if, instead, they made students study writing? How to compose a clear sentence, and how to write intriguingly. If academics wrote more clearly, would they think more clearly? I bet they would. You should've seen my thinking before I started writing.

Awful writing is just the tip of the iceberg; the mass of ice itself is the morass of "data." If any discipline needs sound data and logical argumentation, it's one which cannot conduct experiments. Real sciences pinpoint quantities and take measurements quite seriously (maybe because they can). Economists, on the other hand, cite "statistics" that are far from foolproof (next chapter). Hence economists can believe and espouse anything – and do.

CHAPTER 17

THE US FED ALSO RIPS OFFICIAL US STATS

Facts are stubborn things, but statistics are pliable.

– Mark Twain

BE FOREWARNED

Our government bureaucrats gather measurements then massage the numbers to create statistics. Yet from one bureau to the next, the counts vary. It's as if the statisticians force the figures through vigorous calisthenics to make them flexible.

Stepping up to the plate, the US Surgeon General required academic publishers to paste onto articles about economics a caveat, like on a pack of cigarettes: *"Warning: Statistics contain high levels of political influence. Trusting them can be hazardous to your financial health."* And if he didn't, he should.

Some insiders find official figures worse than useless. Most professional economists don't mind the absence of a total for the worth of Earth in America but do object to the presence of tallies that are inaccurate and irrelevant and thus distracting. Official fuzzy numbers are a perfect match for the standard risk-adverse jargon (Ch 16).

Experts cite them, and most people base crucial decisions on them. Especially when the figures come couched in officialese, people tend to accept them as gospel. But are they? Are they tainted by politics? Is a competing unofficial number better? Decider, beware. One should take these figures with a grain of salt.

At the risk of biting the hand that feeds us, and kicking a dog while it's down, and mixing metaphors, let's continue the critique. Not that we love finding fault, it's just that we love knowing facts. Conversely, officialdom seems unable or unwilling to convey them.

INACCURACY

One expects criticism of official statistics from critics of either left or right – but check these out; sometimes the critics of statistics are mainstreamers.

"...the focus on the two headline indicators ... created great incentives to governments to compile figures on deficit and debt that look good, instead of them being good from an economic substance POV. There is a clear tendency to continuously look for 'grey areas' to manipulate the relevant national accounts data... These practices have substantially increased in 'popularity' since the start of the financial crisis during which significant pressures on government finance emerged, amongst others by the direct and indirect effects of the economic downturn and the bailouts of banks."
    – "Government Finance Indicators: Truth and Myth"

The OECD has produced useful reports before on the link between land value and economic growth, so down the road maybe they could become a standard bearer for determining the size of all locational value.

If only US public agencies had the chutzpah to cry out when the emperor wears no clothes. Well, actually, sometimes some bureaucrats do. Officials at the US Federal Reserve and their staff are already dismissing large swathes of the most recent economic data because they view it as unreliable.

Economic data is constantly revised, and final reads are often significantly higher or lower than initial measurements. Twisting around the stats can leave investors, businesses, and households twisting in the wind. Their plans can be wrecked by the central bank's next interest-rate move.

There are three types of lies -- lies, damn lies, and statistics.
    – Benjamin Disraeli

Like an urban myth, groundless numbers persist. Diane B. Paul, formerly an associate professor of political science at the University of Massachusetts, wrote a book about that: *The Nine Lives of Discredited Data*. Once entrenched, false figures escape detection and hence correction.

Then economists who play it safe – safety is where the money and honors are – perform calculations using approved numbers. Thereby GIGO (Garbage in, Garbage out) strikes again. According to Otis Dudley Duncan (1921-2004) in Notes on Social Measurement: Historical and Critical, those academics suffer from statisticism.

"Coupled with downright incompetence in statistics, we often find the syndrome that I have come to call statisticism: the notion that computing is synonymous with doing research, the naïve faith that statistics is a complete or sufficient basis for scientific methodology, the superstition that statistical formulas exist for evaluating such things as the relative merits of different substantive theories or the 'importance' of the causes of a 'dependent variable'; and the delusion that decomposing the co-variations of some arbitrary and haphazardly assembled collection of variables can somehow justify not only a 'causal model' but also, praise a mark, a 'measurement model'. There would be no point in deploring such caricatures of the scientific enterprise if there were a clearly identifiable sector of social science research wherein such fallacies were clearly recognized and [kept] emphatically out of bounds."

ALTERNATIVES

While our public agencies do not tell us how much we're all spending for the land and nature we use in total, let's not feel singled out. They slight other curious groups, too, who'd like to know statistics like a qualitative GDP, the true inflation rate, real unemployment rate, total assets of governments, actual debts of governments, etc. The more important the indicator, the more massaging it gets.

"There are three sorts of economist. Those who can count, and those who can't."

• Ecological economists object to GDP since it measures quantity of growth, not quality of growth. E.g., clear-cutting trees from a hillside, causing erosion that degrades a stream, contributes to GDP no differently than does selective logging that leaves a forest available to hunters and hikers. Nevertheless, the media report faster growth – no matter what kind – as a social good. And whoever is in office gladly takes credit for it.

The Report by the Commission on the Measurement of Economic Performance and Social Progress (Stiglitz et al.) says, "the time is ripe for our measurement system to shift emphasis from measuring economic production to measuring people's well-being."

• Leftists economists point out that the definition of unemployment was changed to consider the under-employed as employed and to not consider those no longer futilely seeking a job as unemployed. A smaller figure for unemployment, in the eyes of many, makes those in office look good. The real unemployment figure is actually double, or triple, or over quintuple the Bureau of Labor's figure.

• Populist economists remind anyone listening that the official definition of inflation has been changed at least 20 times in 30 years. One of those official changes deleted the very thing we're looking for, which is the value of locations. Using the older definition, inflation would be at least 7%, probably more like 10%.

These critics come from within the discipline, so they themselves don't lack credibility. Their alternative stats do. What they gain in accuracy, they lose in credibility. The alternatives rely on raw data that come from officialdom. And even if unofficial calculators can find a way around that conundrum, their measure is still not official. Hence, nobody pays it much attention. The major players making policy and huge investments ignore the homemade figures and stick with convention.

While I feel for all those critics cited above being ignored by most of their colleagues, their suffering is soothing. Their voices shouting in the statistical wilderness give us room to talk, to question the official dismissal of rents. As we search for any sign of rents, the relevant stats we've found do not inspire gobs of confidence. It feels better knowing other critics have gone before.

While the above academics were able to fault an existing stat, they have their own axe to grind and their own pet theory to promote. They leave themselves open to a different criticism. They were not able to critique the absence of a statistic – the worth of Earth.

IRRELEVANCE

No matter what phenomenon they measure, bureaucracies always fail to agree on one estimate. The statisticians of one bureaucracy cannot explain why the totals of another bureaucracy differ. Nor do they seem to take these discrepancies seriously.

Most professional economists shrug off absent data, even misleading "data." Too many academics are indifferent. They have an attitude of "oh, that's good enough" when clearly the figures are not. It's like they and their statistician brethren have jobs with no curiosity allowed. Caution and conformity should be the job requirements listed right under the job title for a Public Information Officer.

For the academics and bureaucrats compiling them, the jumble of tables is what's important. Whether they have any accuracy or utility or insight does not seem to matter. Doing a job that pays well and gets paid attention from the business media and academics authoring articles (since officials have a monopoly on both data and status, where else can the curious turn?), that's what matters, not the datum for Earth's worth.

In particular, bureaucracies ... count things of minor import – e.g., consumer confidence; over-count some indicators – e.g., GDP;under-count other indicators – e.g., unemployment or inflation; and bundle what should be kept apart – e.g., housing with utilities. Their false frames yield misunderstandings and distorted world views.

Statisticians go deep but not broad. Going deep, economic statisticians measure an enormous quantity of minutiae, like "advance US retail and food services sales." Failing to go broad, they leave out customs like trust which make civilized trade possible. Alan Greenspan, who was the most powerful person in economics while he reigned at the Federal Reserve, confessed to being surprised to learn that trust matters. That was his comment upon observing Russian criminals take over the conversion of so-called communism into capitalism.

Conversely, economists go broad but not deep. Despite economies being nothing if not systems of incentives, economists fail to unbundle the two kinds of spending. Spending for human-made goods and services versus for natural assets are as different as a beard and a barbarian. The economists's catch-all category for spending is reminiscent of speed-reader Woody Allen's review of Tolstoy's *War and Peace* (or *War and the World* in the author's native language): "It was about some Russians." It was about some purchases.

And to top it off, going too broad, economists include political behavior, like lobbying within market behavior like producing output. They fault "market failure" when actually what happened was "lobbying success," or in their jargon, successful "rent-seeking."

RENT-LESS WONDERS

Official stats are not only way off the mark – real GDP is lower, real inflation higher – but their measurements shed little light on what their measurements shed little light on what economies are up to. And what we need to do in response. Not knowing how much society spends to never reward labor and and capital (i.e., our spending for land) means that economists cannot make good statistical arguments. That guarantees the futility of economics.

Trying to calculate aggregates of items is just the opposite of measuring the size of particles. Physicists have their angstroms down to the trillionth. Chemists measure parts per billion. Economists pretend that their stats are of equivalent stature and call their statistics "data." No way. The commonly applied phrase "massaged data" is only half right; there is no data on the menu – only approximations. An official figure resembles an actual value about as much as a stick figure resembles a living body.

Are bad stats mistakes or conscious incorrections? It's like those guys go out of their way to not make sense. Official figures create a near impenetrable fog that hides rent – a useful smokescreen for somebody's capturing of rents.

What officials attempt – aggregating many sales into one grand total – is challenging. Even under the best of conditions, as prices are always fluctuating, it's not easy. But add the political pressure to look away from land and you get the mess we got.

At the end of the day, it's a lot of noise to go with precious little signal. Official tabulation is a morass and it gets worse. Just wait until we reveal where the official statistics for housing and other proxies for land went astray.

CHAPTER 18

FIGURES FOR LAND: FROM A FUN HOUSE?

"Studies show: 79.48% of all statistics are made up on the spot.
                                                — John A. Paulos"

OTHER TOTALS ARE IFFY BUT AT LEAST EXIST

You saw how academic jargon (Ch 16) and official statistics (Ch 17) in general are tough-sledding. Now hang on to your hat. The special treatment economists and bureaucrats give the worth of Earth in America is worse.

First of all, a total for the value of American land and resources does not exist. Not one public site, from local to federal has a separate category for the total value of the nature we use. Researchers must ferret out surrogates. Then their numbers that do exist typically minimize the value of a part of the natural world humans use. For some reason.

That officials don't tally a figure for payments for land, a basic factor, begs an explanation. They tally figures for payments for the other two factors in production, labor and capital. Aggregate wages were $6.3 trillion in 2015. A stat from which to derive interests says that all US businesses grossed about $20 trillion in 2012. Yet officials give no approximate stat for the base factor, land.

We could use one. With an accurate tally, we could fend off this economy that hurts too many people. A total for rent – how much we all together spend to occupy land – would inform us about two crucial phenomena. We'd see how well the economy is doing, plus where the economy is headed.

That the average Jane and Joe overlook land, hidden beneath buildings, is understandable. Us non-experts may refer to the cost of housing, assuming the human-made part to be what's valuable and the nature-made part not so much. But the "expert"? Despite their training – or because of it – even specialists make the same mistake.

CASTING PEARLS BEFORE SWAINS OF DATA

But not all. Some economists can't resist trying to figure out the value of land. Recall Morris A. Davis (U Wisconsin) and Michael G.

Palumbo (Federal Reserve Board) in their "The Price of Residential Land in Large U.S. Cities" (Ch 13). In 2004, two years before the peak in "home" (actually, home plus site) prices, the prices of homes along the coasts were not much greater than those for the other three US regions. Thus the increase in homes (+site) prices reflected the value of land, i.e., location.

The average lot sold for about $440,000 in West Coast cities. Land's share of home value had risen to 75% along the West Coast. When sampling upscale neighborhoods of cities, that figure easily rises to 80% (once again, the *Pareto Optimum*) and even more of property value.

Land's share of home value had risen to 65% on the East Coast, compared with about 40% in the other 3 regions. And here's the key stat: it rose 51% across the entire sample of 46 cities.

"50% of marriages end in divorce. Thus, if you don't file for divorce, your wife will."

Usually, when bureaucrats and academics put forth a number, it's lower than what's logical, typically lower than how much private business calculates. In the business press, one can read the occasional article reminding the lay public that location absorbs purchasing power. Hence land is vastly more valuable than buildings. One article found this obvious yet invisible fact in the 10 most populous US cities.

Davis and Palumbo aren't the only game researchers. The Lincoln Institute calculates land value based on home sales and updates it frequently. Eschewing houses, Albouy, turned up $30 trillion for metro land, while leaving out the trillions for non-metro turf. Larson took a stab at the grand total and found $23 trillion, way below Albouy's partial total (Ch 13).

THEY COUNT *SOMETHING* – BUT WHAT?

While grateful for these exceptions, the rule is another matter. Not only do statisticians lack a stat for rent, they also christen their charts for the profit from location with jargony labels and sometimes without any labels at all. To get to the meaning of their numbers within, one must decipher their packaging.

Contact an authoritative voice: the Census Bureau (public) or the Federal Reserve (private). One of the surprises among many is that authorities did not understand my question: How much do we spend for the nature we use each year, from land to oil, including water, etc? Such curiosity must be thinking way outside their box.

Yet at first, the "experts" are flattered that a lay person takes their field seriously. But soon an investigator's determination to pry an answer out of them begins to seem a fixation. Their recourse is never rudeness; instead, it's stonewalling. "Nothing to see here, move along."

Are they right? Are all matters economic better left to the experts? Driving away us riffraff, is that the official rationale behind the present figures that are incompressible and irrelevant? But if we don't persist, who will? Inside the official box, land is obscured, subsumed under capital; i. e., buildings. Bureaucrats suggest, *If you want to see something relevant, look at what stats we do trot out!*

Eventually a determined persistent one does decode the jargon, but why should one have to? The public pays for the gathering of the figures. Shades of Paul Romer (Ch 16), the paid gatherers should present their findings in a format as accessible as possible.

Beyond appearance is content. Despite missing much rent, specialists go ahead and publish official or scholarly underestimates. These statistics you do find...

- come based on assessments or appraisals of lump-sum sales, not ongoing leases;

- may combine commercial and residential and leave out agricultural;

- may include sylvan, mineral, spectral – or not;

- tally together private and public – or not;

- miss elements such as roadways, port districts, airport landings slots, the value of marina slips, etc; and

- are not centralized or standardized but are scattered across many agencies, both public and private, and in language that is not only obtuse but inconsistent from one agency to another; they don't use the same meanings for the same terms.

Most tabulators don't give a value for land but for land and buildings combined. Those few who sought the value of land alone tried to separate it from the price of location plus house. Yet the worth of Earth is far greater than the value of just the land beneath a single-family home. Downtown – land beneath stores and offices – is where site values spike. And the method they chose to ferret out land under-counted it.

Replacement Cost? Land Costs Zero!

To appear smaller, dress the stat in horizontal lines.

Conventional economists claim the value of buildings far outweighs the value of the location. Trying to justify their assertion, they don't subtract the value of the house as is from the combined total of the property. Instead, they subtract the building's replacement value, pegging the value of that house as if it were brand new.

Their method is like saying the value of a driveway with a junker resting on cinder blocks is not the value of the driveway by itself. Instead, it equals the value of both combined minus the value of a brand new Camry. Those specialists could use a BlueBook for used houses.

Want to know the actual worth of your home? Put it on a 30 foot wide trailer, drag it around town, and see what offers you get. That'll be built value. Subtract *that* from the combined value. Your remainder will be the value of the location.

If insisting upon using the method of replacement cost, use the replacement cost of the land. Use a fraction of the $166 trillions that Costanza tabulated for the world's ecosystems (Ch 4). A *reducto ad absurdum*?

Buildings age, need repairs; they depreciate (they're not known as "money pits" for nothing). Regions, on the other hand, become popular and more densely settled. When new people move in – like high-paid techies into San Francisco, or sold-out Californians into Oregon – locations appreciate. The pushed-up property values are pure land value, nothing to do with aging buildings.

Bureaucrats make available a morass of surrogate numbers woefully distorted. It's like economists go out of their way to issue under-sized values for land or locations. If they're not going to do it right, why do it at all? Imagine that the local pub's barkeep served residents shots so shaved down. I'll bet then the public would be up in arms.

Academics, bureaucrats, and dynasties (to coin a term) studiously ignore our spending that never rewards labor or capital (since neither factor created land). They subsume rent payments in expenditures for goods and services that humans do provide, then release their figures weekly, busily turning lemonade into lemons. Filling the forum for knowledge with noise makes it a great place to hide the rent signal. Out of sight …

EYES WIDE SHUT

Those officials who collect the figures have neglected rent for as long as they have been collecting figures. Similarly, economists have mislaid land for so long, that today's cloistered practitioners fail to see the non-human factor's relevance. If land does not count with economists, why would statisticians count its value?

Economists do not demand relevant and accurate data, and public servants do not supply them. Instead, both make excuses for each other. *"Determining the value of land is difficult."* And, *"Such a stat offers little utility anyway."*

By now, this official trivializing of rents has bred indifference among conventional economists. For them, society's spending for one of the three factors in production – land – does not matter. The vast majority of economists simply do not care how much it (locations, natural resources, EM spectrum, et al) is worth. After a while, they can quit even pretending to tabulate land value. Rather than calculate accurately, the Fed quit calculating rent at all, notes the OECD.

The official gap in data for land and rent blinds people. Laypeople conclude, logically, that the items that officials do count matter and the items they don't count don't. Due in part to this negligence, natural values and occupied land have become invisible to the modern naked eye.

Professionals who should know better don't see the difference between assets created by humans and those that are not. Established economists don't see how our spending for human-smade assets stimulates more output while our spending for natural assets does not. Hence they miss what drives the business cycle.

POLITICAL PRESSURE

To a degree statisticians are innocent. The laws in some states put caps and limits on property taxes and thus force down the official estimates. Maybe someone is keeping the good data private somewhere for well-connected insiders.

Even politicians are innocent. They're obeying the will of voters, and the most consistent bloc of voters are homeowners. Most of them are equal part land speculator. Dealing with land value puts one under political pressure. Most everyone yields ("Politics and Federal Statistics" by Janet L. Norwood in Statistics and Public Policy, 28 Sep 2016).

Appraisers, being in the employ of realtors, usually round estimates up, as jacked-up appraisals tend to inflame ground traffickers. Furthermore, sellers and buyers do not always convey the exact price to the assessor's office. Sometimes they shave it to lower tax repercussions.

Assessors, when they exaggerate the value of the improvement, they do the property owner a favor, since buildings depreciate and the owner can deduct that amount from their income tax liability. When they understate the value of the location, they do the land speculator a favor, since a lower number hides this socially-generated value.

Such distortions do not go unnoticed by appreciative insiders. Conversely, accuracy brings attention of the unwanted kind, putting one's career on the line. Hence the prevailing political tide carries official figures away from accuracy.

For conventional professionals, a realistic large number is controversial, while a sketchy small number for rent is safe. However, less really *is* less. Controlling the flow of information is a way to censor and to marginalize those who seek answers. That impoverishes culture, since knowing and understanding how one's world works is a huge part of any culture. Specifically, society loses the measure of natural surplus, a reliable indicator of economic swings.

While both right and left economists do well in keeping themselves out of poverty, how well have they performed lifting others out of poverty? Or preventing large swaths of the middle class from falling into poverty when economies shrink and no mainstream economist warned the trusting public?

Perhaps academics and bureaucrats don't see how one spending stream rewards privilege and other streams do not. However, it's more likely they do see the elite being benefited, and that's the problem. Curiosity gives way to caution.

Downplaying the role of land and payments for land, those economists and statisticians stay the course. Or, if you can stand the pun, they stay the curse. They are not doing science but merely maintaining the status quo. Whether intending to or not, without received orders being made explicit, both number-crunchers and academics have become guardians of the rentiers, the few happy recipients of the vast flows of rents.

Or they perpetuate the main rival ideology – any of the various left-isms. Those also overlook land, limiting themselves to the usual labor-versus-capital argument. Leftists reinforce the industrial paradigm and likewise miss the organic nature of economies.

Yet despite pressure, researchers in other fields do science even when opposed. Communist economists have happy careers in academia. Evolutionary biologists, harangued by religious myth makers, tell their truth. Why can't mainstream economists and statisticians do it, too?

ASSESSING ACCURATELY

Have we set the bar too high? No, official American statisticians have set the bar way too low.

Since localities do not tax locations apart from improvements, they don't care about the value of land. Ever the practical, county assessment offices quit assessing nature-made locations separately from human-made improvements. In Oregon, some assessor offices toss into the circular file their assessments of land. And being officials, the way they do things is what becomes Standard Operating Procedure.

Some assessor offices do much better. British Columbia, whose office was set up by geonomists Mason Gaffney, emeritus UC-Riverside, and his protege Ted Gwartney, former assessor for several jurisdictions and Chief Appraiser for Bank of America, is so well-known for precise assessments that professionals come from all over the world to be trained or hone their skills in rainy BC. Americans could reach for the higher Canadian standard.

As for any difficulty of separating the values of land and buildings, the difficult is not the impossible. And how hard is it, really (outside of political pressure)? Actually, not that hard. Redevelopers who buy property to tear down the extant building and erect a new one, they separate the value of the location from the value of the improvement upon it all the time. For them, the task is simple.

With computers, the county assessor's office could easily update all records of all parcels in an area every time one of them sells or leases. The totals would never lag and be available to all comers. The federal agencies, who now get their numbers from the local ones but don't do much with them, could truly serve the public.

The "incomparable John Rutledge" already does. In his "Total Assets of the US Economy $188 Trillion, 13.4xGDP" he notes that the Federal Reserve Board furnishes balance sheets for non-financial assets covering several sectors of the economy. Yet the Fed does not include the value of land and non-produced assets held by financial corporations and government. Moreover, owners do not directly report the value of their land. This gap between concept and measurement biases figures downward.

The link to this article on his website no longer works, but in a private email John estimates the 2017 total to exceed $300 trillion; a portion of that is land price.

LAY DOWN THE GAUNTLET

While accuracy would be great, at this point we can make do with a ballpark figure. Heck, none of the other stats from bureaucrats and academics are precise. Yet they keep cranking them out and passing them off as "data," as if they were milliseconds or parts per billion.

Consider the firepower of all those public agencies spewing reams of numbers that don't draw a picture but just muddy the waters. Imagine if those staffs were directed to measure essentials. It'd be so easy for them to calculate the worth of Earth in America.

There are tons of numbers to use. Each year in the US, there are hundreds of thousands of transactions for land, resources, electromagnetic spectrum, etc. There are sales, leases, sublets, auctions, plus taxes levied on those natural assets, not to mention interest paid on the land portion of mortgages.

Most likely, number-crunchers won't budge until economists ask for more relevant stats. Perhaps it requires a controversy – an innocent proclaiming that the emperor wears no clothes – to motivate the discipline, to spur reform. It's a duty job, so somebody has to do it, typically a gadfly, a curious investigator, certainly not an official obfuscater.

The job of this book is to raise the bar, to raise expectations. How will criticized gatekeepers react to an outsider laying down the gauntlet? Ignore and ontinue to mislead? Or find the chutzpah to ferret out facts?

No hard feelings, but may economists remember, science has always been a back and forth between old and new theories, and between researcher and skeptical public. Once they're measuring the flow of rent, seeing the size and fluctuations of that spending stream, and understanding how economies operate, economists could do more than guesstimate; they could predict. Finally forecasting accurately – and stating dates people could plan by – they'd become real scientists. Thereby, our critiquing of their nowadays numbers is a service to society.

It makes me feel a little sorry for economists and statisticians who once may have had professional pride, their having to toe such a partial, biased line. Do they feel embarrassed to be dishing out such distortions rather than the best data? To be the butt of jokes? (e.g., somebody who rescues statistics that can't lie for themselves). Deep inside, I bet economists probably want to become scientists, or they would not suffer from physics envy. A healthy number of them would welcome reform, eh?

CHAPTER 19

COLLUDING OR NOT – THEY CENSOR RENTS

People who think they know everything are a great annoyance to those of us who do.

– Isaac Asimov

D o too many people want to know the worth of Earth in America and secrets must be kept? Or do not enough people want to know and paltry demand elicits no response? Whichever, those with the resources to measure natural value in America, but don't, have made it exceedingly difficult for those who don't know that value, but want to find it out. In a perversly ironic way, data centers have made it harder than ever for the inquiring public to access what should be public records.

JUST LIKE CENSORSHIP

U sed to be, you could turn to the county cadastre and find the official value of never-produced land (however accurate) apart from human-produced buildings. But not anymore. Economics' drift toward a landless discipline has culminated in a rent-less database.

How can economists and statisticians not burn with curiosity to know how much their society spends on the nature it uses? Alas…

- Many county assessors no longer separate the values of locations and improvements.

- When they do, they do not bother to be precise.

- Some counties don't forward their assessments to states. States, in turn, pass on the deficient stats to the Census Bureau. That bureau, despite being bereft of all counts, is the source for the Bureau of Economic Analysis, the Uber Bank – the US Federal Reserve – *et al.*

When local assessors quit giving a total for land, so did the Census Bureau. And so did the pseudo public Federal Reserve. And so did the truly private Zillow. Tuning out society's spending for land was gradual, then worsened. Now it's complete.

Public agencies aren't quite as public as they used to be.

- States contract out penitentiaries.

- The army hires mercenaries.

- The Senate and House of Representatives debated privatizing the Federal Aviation Agency.

- Media commentators proposed selling off the TSA and the Post Office.

- Congress gave the power to issue new money to a consortium of private banks. And ...

- They, Federal Reserve, outsources to private corporations for the statistics that they themselves used to collect, collate, and package in quasi-intelligible form (Ch 26).

If you ask, the Fed introduces you to the private firm that they use. If you want those values for the surplus the economy generates from natural inputs, you have to pay. For the middle man Fed, it's pennies. But for your average quester on a shoe-string budget, it's more than a researcher can typically afford.

Another user of public assessor numbers, the private Lincoln Institute, after all these years, still won't expand their total from residential land to all land in any use.

Turn from public assessors to private appraisers. Bankers accept property as collateral. When they foreclose, they resell the land plus building. Eventually their borrowers resell or refinance. As sound business people, bankers might like to anticipate their prospects – how much profit they can expect in the future from the grand total of all land.

The American Bankers Association has the statistics. Until recently, a researcher could visit or ask the librarian for answers. But no more; the ABA closed their library along with its research service.

This official tuning out of Earth's worth has been a gradual transition. It just got really bad lately. Now it's complete.

Nothing Personal

The number-crunchers' failure to meet the needs of we total-doters is not to be taken personally. Being mighty, those specialists need not bother with rebuffing a curious gadfly. Indeed, given the influence of rentiers, it was inevitable.

Getting *any* response assumes you actually reach an official statistician. *"Hello, can you tell me the worth of Earth in America?"* Usually it's like calling a big corporation – a runaround. You get transferred so many times that ultimately you end up talking to the same person you started with.

Public servants in charge of public information don't have to serve the public useful answers.

- They can give unrelated facts and figures.
- They can say the total of land value is not known and is not knowable. And...
- They can just not reply to requests at all.

It's the bureaucratic version of W.C. Fields', "Go away kid, you bother me." Even longer ago, having to dish out such futility at his job is what supposedly drove bureaucrat and absurdist writer Franz Kafka, author of *Metamorphosis*, mad. If their goal was to utterly derail any inquiry into the value of land and resources, officials have come close.

The indifference of established academics dampens any interest of newcomers, who could become potential researchers. Which is sad, because breakthroughs usually come from newcomers. That leaves turning economics into a science up to outsiders.

Face it. Statisticians are a speed bump; economists a road block. The elite benefactors are a dead end. So, Do it Yourself?

No problem. We'll make the most sense possible out of the best numbers available. And conjure a total that perhaps the specialists will stoop to critique.

CHAPTER 20

True Total Trillions Yanks Spend on Land

If you torture data enough, it will confess.

He Asked for It

Bill Larson, an employee of the US Bureau of Economic Analysis, offered the world a total for the worth of Earth in America. Now that others can see his figure for the value of land and resources, that rent becomes real – and useful. A serviceable estimate for how much we spend on the nature we use fills in a huge gap in our numerical knowledge of economies. The statistic reveals the size of the surplus that society generates, plus, when paired with income, what phase of the business cycle we're in.

Bill sought the entire value of natural assets, urban and rural. He did his research and calculations on his own. For whatever reason, his and all the other US agencies do not provide such a number.

Bill's tally came to $23 trillion for all land. However, that's $7 trillion shy of Albouy's total for metro land alone, excluding natural resources (Ch 13). What explains the wide disparity?

When including land owned by state and local governments, did Bill include roadways? Most overlook streets, but not all. TeleCommUnity in 2002 figured the annual value of the nation's rights-of-way by themselves to range between $305 billion and $366 billion, and the price of the land beneath streets – most streets are in high-value cities – to be $7.1 trillion[1] – the difference between his $23 t and Albouy's $30t.

Toss farmland and everything from the subsurface like oil to the supra-surface like airwaves onto Alboury's $30 trill, and Bill's figure looks $17t short (Ch 15); for us, a trillion is still real money.

Earlier scholars used figures almost a decade old; let's use up-to-date totals. Most researchers tabulated lump sum prices; we'll calculate more precise rents. They all left out something; we'll corral all natural assets.

---

1   "Roadway Land Value" Page 5.7-18 in Transportation Cost and Benefit Analysis II by Todd Litman at Victoria Transport Policy Institute, 21 March 2019

Larson called his figure a draft estimate, said his study attempted to provide baseline methods upon which further researchers can draw. He asked fellow economists and statisticians for refinement, yet his brethren ignored him. We won't. If public agencies won't do the work, we'll have to figure it out ourselves. It'd be great to have an army of eager grad students pitch in. Professors have that luxury, so why don't they … never mind. If you want something done right…

What we lack in experience, we'll make up for in unbiased adherence to impartial logic. A real scientist goes out on a limb and says the world is round, space is flexible, time relative, and things too small for you to see make you get sick and die, no matter the fuss it may kick up. Political prudence and compromise is nowhere near doing real science. Unlike those whose careers are on the line, we can follow reason wherever it may lead. So we won't low- or high-ball. We'll merely tally as precisely as possible. And live with the results.

To total up all the rents, one needs the figures for all:

- Land: downtowns, home sites, farms, fields, forests, underground ores, oil deposits, etc,

- Water rights and waterways,

- Airwaves for all telecommunications, and

- And if you're not to leave anything out, incidental locations like marina berths, airport landing-lots, etc,

- The above, whether owned privately or publicly, and

- payments of principle, interest, insurance, taxes, etc on these assets. That should cover it.

HE'S GETTING WARM

The one who generated a total for all land, Larson, admitted his was a best guess, one with gaps that needed filling. When including land owned by state and local governments, did he include roadways? By focusing only on the surface, could he encompass oil, ores, and airwaves? Did Larson incorporate the "F" and "I" of F.I.R.E. (payments for loans and insurance)? The answers, as far as I could tell, are "no."

The biggest total – $30 trillion by Albouy – was also incomplete, for metro land alone. Yet it clobbered all other academic estimates for *total* property – natural land plus man-made buildings. The greatest amount for land any one of them could find was $12 trillion (Ch 13). How can the discrepancy be so great?

- First, most researchers examined only its most common use, residential land, leaving out its most pricey use, commercial land.

- Second, when officials subtracted a value for buildings, they subtracted a replacement cost. Doing that ignores depreciation of the building, exaggerates its value, and shrivels the value of the location. Larson used those shriveled estimates for residential land.

- Third, he also used official figures for land in other uses, when he could find them. Those figures were woefully incomplete and similarly minimized (Ch 18). For some reason, official figures, even Larson's, are gross underestimates for some reason.

- Fourth and fifth, Larson unavoidably left out lots of things and used figures for 2009, close to the bottom of land prices in the recent recession. Albouy, too, left out lots – all rural land – and also based his total on prices from 2009. What Albouy did differently was avoid aggregate assessments of land plus buildings (Ch 13).

Instead, Albouy's team used individual sales. Not sales of property but only sales of land. From those prices, they extrapolated a total for all land. Just as "replacement cost" and other tricks make official small figures unrealistic, actual land sales make Albouy's big figure reliably realistic, despite being so much greater.

FILLING IN THE BLANKS

Caveats: I'm able to use only the publicly available stats. Maybe the stats one pays a business for are better. Maybe somewhere are good stats for free. So it goes…. We'll play with the cards we're dealt.

Since Albouy published (Ch 13), housing has recovered all its lost value and more. Because most of the change in housing value is not in the building but in its location component, we can use that percentage[2] increase for land.

According to Zillow, since 2009, the increase to 2018 has been 12%. So Albouy's $30 trillion is likely in 2018 to be $33.6 tr. Coincidentally, the increase in federal land since 2015 – the year of Ebeling's $5.5 tr (Ch 15) – has likewise been 12%. That raises federal holdings to $6.6 tr – more than the $6.5 tr tax dollars gone missing. Add up these two updates of Albouy and Ebeling, plus the $3 tr for farmland – the USDA figure was up-to-date (Ch 15) – plus the fraction for water. Now we're at $43.5 trillion.

---

2   "Why Have Housing Prices Gone Up?" by Edward L. Glaeser, Joseph Gyourko, & Raven E. Saks; 2005

Forty-three and a half trillion for all land and natural resources. After spending so much time down the rabbit hole of officialdom, I feel like I'm performing alchemy, cobbling together such a biggie when they arrive at such littles. Besting the best official guesstimate by so much does give one pause.

Now that we've brought price up to date, let's bring it down to earth and convert price to rent, before we seek values for all parts of the natural world we use.

FROM PRICE TO RENT

Conventional economists mix apples and oranges. They compare the price of land, a lump sum, to GDP, a flow. If comparing to GDP, they should cite rent, another flow. If citing price, they should compare to other fixed assets, like buildings. Apples to apples, oranges to oranges.

Steve Cord, using National Income (Ch 13), and Riley Ashton, using leases (Ch 11), calculated the annual rental value of all land. They had good reason to. Rent is fundamental, price is derived. To charge rent, owners don't have to know price. Yet to set a price, first someone must know the rent.

To evaluate sites, one measures annual output. Not just farmland, which we judge by harvests, but commercial sites, too; we assess how much merchants make there yearly. Even homesites have benefits that are seasonal—a view of nature, proximity to a school or a park, etc.

Then one estimates a decade-or-more's worth of rent to set a price. Being a guess, price has a builtin chance of being inaccurate. Being an actual payment (usually), rent has builtin accuracy.

While an ongoing rental value is a fact, a lump sum payment for all land is a fiction. If all land were for sale, it could never sell for the aggregate sale price. Supply would drown demand. Auctioneers could never fetch anywhere near those aggrandized aggregates. With so much to choose from, buyers could be picky and patient. They could negotiate down to an amount well below lump price, closer to annual value.

If society leased out all locations at once – a la Hong Kong, Canberra, US port districts, et al – yes, society could recover land's rental value every month. Tenants could afford periodic payments. Leasing being feasible, rent totals tell a truer story. Rent, not price, is what measures the value of land.

Ironically, to derive an aggregate rent, the process goes: Owners push rent to the limit, from that sellers set prices, from that specialists take a ratio of price to rent, divide price by it, and estimate a new number for rent—which everybody hopes equals the original amount. Fred Foldvary and Tim Worstall (Ch 13) sought their grand totals that way

Some say the ratio of price to rent is the rate one usually paid to borrow not land but money: 5% or 20:1. Yet throughout history, the relentless competition between unequal economic agents fluctuated, and with it the lending rate. Twice that common figure – 10% or 10:1 – is realistic, too. Look:

- HUD's average selling price of apartment complexes (multi-family housing) was $1 million. Their annual average of rental receipts was $100k, one tenth. That ratio is about 10:1.

- The OMB's estimate of federal receipts from federal land was about $0.5 trillion. Ebeling's estimated price for federal natural assets was $5.5t. Again the ratio is about 10:1.

- The accumulated price for land from Zillow, Lincoln, and the BEA were about $27t-$28t. At that time, Quora figured we spend at least $2.72t on land (at least because Quora said the data they found had gaps). Once more, the 10 to 1 ratio holds.

Applying the 10:1 ratio to our own estimate of the aggregate price of land puts land rent at $4.35 trillion. Yet we're not finished. People don't just pay a price or rent to own or control some land or resource. They also pay interest, insurance, taxes, etc for land. *Census says the property tax was over $503 billion in 2016.* Since then *property has increased 43%, says Zillow,* so property tax payments must by now exceed a half billion. So rounding up to $4.4 trillion is easily a safe bet.

If you can't count on their counts, then extrapolate the way a geonomist does.

GEONOMISTS TAKE THEIR TURNS

If you think $4.4t is a lot, check out what our friends with credentials have estimated for the total of all rents:

- Dr Nic Tideman, a former Presidential Advisor on inflation, calculated a ballpark lump sum price of all land at $31 trillion (private email). He put land rent at roughly 10% of GDP or NDP. That's about $2t, a far cry from his $31t, and neither 5% nor 10% of his aggregate price. Go figure.

- However, Dr Steven Cord found rent to account for a quarter of national income and our $4.4t is about a quarter of current national income.

- Dr Mason Gaffney of UC Riverside put the figure for society's spending on land and resources at $5.3 trillion.

- Dr Fred Foldvary at San Jose State chose a third of National Income – land being a third of the main factors – equaling $6 trillion that year.

How can two of these totals be trillions more than mine? Mine left out portions of rural land value; in the countryside, there is more land than just farmland. And there is more "land" underground – oil and ores. Mason may have included what I could not find. Fred followed the lead of Dr Terry Dwyer (formerly at Harvard), an Australian who specializes in land calculations and used Aussie figures which are more reliable than US figures.

A total for rent of five or six trillion, with a ratio of 10:1, puts the aggregate price of land at $53 or $60 trillion. That makes our aggregate price for land – $44t – seem not so huge after all. Whether $44t or $60t, all are easily feasible if John Rutledge's $300 trillion for assets (Ch 18) is reliable; land would clock in at $180t, minimum.

Which estimate is the best guess? Given that:

- land is one of the three basic factors, its rent should be a lot,

- Aussie figures are far more accurate than US statistics,

- real estate typically dwarfs stocks and bonds,[3] and

- Rutledge's $300t suggests land alone should be immense, therefore.... These are persuasive reasons to go with *not* my $4.4t but Fred's $6 trillion.

GETTING PERSONAL

All these estimates by geonomists need to be authenticated, but by whom? Critiques can't rest on solid data, because they don't exist. While we welcome critiques, a critic must employ reason and address methodology. That takes experience in dealing with rents, something most economists lack.

Foldvary extrapolated from the experience of an expert, Dwyer. But a sum as big as $6t triggers the old, "If something seems too good to be true ..." Can this be a mirage? Giving into doubters and relying on Albouy's $30t as our base, let's go with my $4.4 trillion, a decent ballpark figure. Per capita, that's $2k/mnth or $24k annually. May this finding nudge statisticians and academics to refine our work – with great ease, since my text isn't in jargon. Then we'll compare my new standard to the estimates of mainstream economists and bureaucrats who'll come later. Kicking off a discussion on rents would be progress, too.

---

3   "Savills World Research: How Much Is The World Worth?" At McGuire Real Estate, on 24 April 2017

CHAPTER 21

HOW DID AN OLD GUARD
TREAT A YOUNG UPSTART?

I asked a statistician for her phone number ... she gave me an estimate.

MORE RENT, FEWER ADMIRERS?

Recall those four plus trillions of dollars we just deduced for the worth of Earth in America? Incredible, eh? Am I pioneer? Or am I walking the plank above watery ignominy? Announce that all we spend on all the nature we use – from oil to atmosphere and every location in between – is grand. Some don't want to hear it, like it's bad news. What, they wanted society's surplus to be small?

Yet it's understandable. Accustomed to the value of land and resources being tiny, then the bigger rent is, the harder it is for conventional economists to believe it – and admit how wrong they've been for so long. Thus the number of appreciators is inversely promotional to the actual quantity of rent.

Other big numbers sure have it easier:

- a statistician says the national debt is trillions of dollars heavy,

- an astronomer says the universe is billions of light years old and big,

- a physicist says a string is a zillionth of subatomic particle,

And you don't blink. But say the value of natural assets is immense ...

While economists may feel physics envy, they don't exhibit physics receptivity to outlandish imagination. Physicists suggest new explanations, no matter how bizarre – like two-way time travel – and even accept such hypotheses. After all, imagination is the sources of new theories.

Conformist economists, though, must first bow to which way the wind is blowing, on behalf of their job security. It's only logical that only a few of them would stick their necks out this early in the game, while most of them would ignore or criticize measuring the value of the factor in production that nobody created.

Admiration – a Two-Way Street

Do specialists who gather economic statistics appreciate a gadfly's effort to tabulate the value that society generates? Happily for me, some official economists and statisticians did not remain on the sidelines but took an interest and lent a hand. Even though my estimate was "not invented here," our quest did inspire some appreciators among the insiders. Some not only replied to inquiries and replied helpfully but often did so with casual humor. That makes number-crunching less burden and more fun.

Without revealing their identities, I'll treat you to a few excerpts from some of my correspondents' replies...

Jeff,
Was it Truman who was so exasperated with his economic advisor always telling him, 'Well, on the other hand...,' that he wanted a one handed economist? There really are different uses for the various surveys! Try asking the Bureau of Economic Analysis (BEA) who bought what!
– Revi K., Bureau of Labor Statistics

Jeff,
How much is land value? It's a lot! (No pun intended :) It's my understanding that the BEA (the people that do the GDP) is working on including land in the National Balance Sheet, but they're not there yet. You would want to clarify whether you are talking just about "urban land" (that is, "real estate" land), or also land that is productive as such in its own right (without buildings) such as farmland, timberland, mining land. Also about whether to include dedicated open space (such as parks – how much is Central Park 'worth'?).

I suggest you contact Professor MD. He subscribes to a data service in order to update his estimate just about daily. [Ed. Note: And the good prof did come through, writing me that all residential land was now priced at only $9 trillion – a pitiful amount for all home sites in America, actually.]
– DGL at MIT"

"Jeff,

Thanks for the detailed comments.

The data is [sic] based on transactions, so it does not include all properties and land in the United States. Extrapolating from this data to the entire nation is possible, but any such measure should come with the appropriate caveats.

I have had discussions with staff at Census on using the data to improve our estimate of land values, but we have not had a chance to work on that project yet.

I am in the process of updating the estimates and hope to be able to publically [sic] release the updated indices by later this summer. I'll be sure to send you a link to the new indices once they are released.

– Jay N., the US Federal Reserve"

"Dear Jeffery,

Would you be available on Friday 4/22 at 2pm for a phone conversation? RL, an economist who follows real estate for the Financial Accounts, can join us then.

Regards,

– Luz Q., the US Federal Reserve"

"Jeffery,

Thanks! I'm glad you find my posts interesting and I really appreciate your note. My experience suggests we haven't identified the effect of population growth, apart from economic growth, on housing prices.

Thanks for sharing that paper. I was happy to see it came from the University of Kentucky. I did my undergraduate there.

– Len K., Freddie Mac"

"Hi Jeffery,

I don't know why the Colorado legislature curtailed municipalities from taking advantage of the real estate transfer tax after Aspen adopted it to use for affordable housing. Two people that I think would be more well versed on the subject would be Mick I. and Rachel R., both of home have been heavily involved in local policy since at least the 1990s. Let me know if you need their contact information.

– Catherine L., Colorado reporter"

Thank you! I too am occasionally bewildered when academics don't create a consistent time series out of an index that could be consistently updated for just a fraction of the original work they put into it. Particularly because it guarantees them a steady stream of citations from people like me (see the Economic Uncertainty Index, for example) and, in some cases, even a series on FRED that untold researchers will end up using and citing. Zillow Research has the best home-value index, because they adjust it for the mix of homes in the area – it's not distorted by the specific mix of homes on the market at any given time as other home-value indices often are.

– Andrew V, *The Post*

Hi Jeffery,

Thanks for writing. Current numbers add up to more like $300T. The best source is the Fed's Z1 report. Link and file below. Near the front you will find a page that lists total financial assets and liabilities bisector and total but beware as they include no tangible assets. Toward the end there are pages for balance sheets but they only cover 3 sectors – including households but not governments. In the household numbers you will see the assets that include tangible assets and report land and structures separately.

Best,

– John R, columnist

Hi. This table is part of the Z.1 release: https://www.federalreserve.gov/releases/z1/ current/html/default.htm. You should be able to find what you are looking for there.

Thank you for your interest in the Z.1.

– Virginia L, Sr Financial Analyst, Board of Governors of FRS

Hi Jeffery,

CBO takes its estimates of the overall value of land in the nonfarm business sector from the Bureau of Labor Statistics. We don't include an estimate of the value of farmland. A benchmark for nonfarm land is estimated by applying a land-structure ratio based on unpublished estimates by the BLS to the value of structures. These ratios are based on data from 2001 for Ohio. This benchmark is extrapolated from Bureau of Economic Analysis investment data. The resulting nonfarm land data series is allocated to industries based on Internal Revenue Service data on book values of land. Land is assumed not to depreciate. I hope that helps. Regards,

– Bob A, Unit Chief, Congressional Budget Office"

Thank you for your question. The government does not determine the fair market value or appraised value for every piece of land the Federal government owns. Funds received for various rights (air, mineral, water, etc.) to real property, agencies may track those funds in their various internal systems, but it is not reported to the FRPP (Federal Real Property Asset Management) system I manage.

– Chris C, US General Services Administration"

MORE RENT ON THE HORIZON

These helpful respondents above represent the cutting edge within the discipline, an edge that should keep on moving forward. As other researchers come sniffing around, looking for reliable totals from official data gatherers, requesting a more rational methodology in their parsing of stats, then professionals such as the above are likely to deliver a finely honed total. Finally we'd have a number we could utilize to predict reliably and clarify how economies really work.

Later, being better informed, with better ways to measure it, experts might expect the assessment of the amount of rent to total even higher. Tabulators would take account of more sources, like marina berths and note that a lot of interest on loans is actually rent. Putting it all together, that's a scary thought for those who now insist the size of all rents is negligible.

For doubters, the size of rent is going to get a lot worse. Beyond land, people pay big bucks for some land-like things, other never-produced assets: airwaves, ecosystem services, and the field of knowledge. We have yet to count those three huge categories of spending. Let's get to it.

CHAPTER 22

MANMADE UTILITIES SPOUT NATURAL RENT

"I'm not denying that monopolies are terrible things. But it is easy to resolve them."
— Alan Greenspan, once the world's most powerful banker

UTILITIES GENERATE RENT, TOO

Over recent decades, inventors have made parts of nature exceedingly valuable, including an invisible part, the electromagnetic spectrum for telecommunication. This ever-advancing technology also makes valuable, along with airwaves, the tangible, human-made cable networks for carrying digitized information. Just as you pay for the rest of useful nature – typically paying rent (technical meaning) for the land beneath your home – you also pay rent for EM frequencies and for cables, strung on roadside poles or buried underground out of sight.

Of course, you don't pay directly, just as you don't directly pay for farmland when you shop at the grocery store. But what you do pay does get passed along. And what you and the rest of us pay in order to telecommunicate, that gets added to the subtotal of what we all pay for the more familiar land. Thereby we get closer to determining how much is the total worth of Earth in America.

While nobody made the EM frequencies, and somebody made the cables, that difference is overshadowed by three similarities.

• First, both are locations – frequencies in the spectrum, cables along the streets.

• Second, both allow monopolies – telecomm corporations and utilities tend to giantism, controlling huge amounts of territory, just like those guys who own millions of acres.

• Third, for those two reasons, whoever owns or controls cable networks or frequencies can charge well above cost – that is, both can charge (technical) rent.

Utilities As Natural

Consider the first point – location. Although human effort is necessary to build utilities, labor is far from sufficient. Just like in real estate, the three most important things in utilities are/is location, location, location in nature: dams need river valleys, railroads need gradual grades, windmills and cell towers need hilltops, satellites need geosynchronous orbits, etc., as cables need rights of way. Otherwise, none of these goods can serve.

Now, the second point – monopoly ownership. Electro-magnetic communication is the latest "natural monopoly." Earlier came the power grid and before that water and sewers. Older still were the railroads, the big bad corporations of their day. The first natural monopoly were streets (below which are the routes for water and sewer). Imagine if different networks in the same region were to compete: parallel streets for cars, parallel cables for communications, parallel sewers for crap – not too efficient. In such cases, monopolies make more sense and are dubbed "natural."

The granddaddy of them all – roads – was almost never owned by a single person. The exceptions were someone owning a bridge or ferry. Or a toll road, which was legal; the illegal variant was highway robbery.

Ownership of today's utilities, water, electricity, etc., has been mixed. Sometimes the public owns, sometimes a private party. The latter, the profit-making utilities, show that these networks generate income – lots of it.

Now, consider the third point, rent. As monopolies, utilities lack competition (duh). Not challenged by someone willing to accept a slimmer profit, whoever owns the utility monopoly is in a position to charge customers a steeper price, one well above cost. *It's something private ones tend to do while public ones don't.*

"Out of sight, out of mind." The airwaves are intangible. And the networks of cables and pipes, though tangible, might as well not be. (For the sake of esthetics, if only telephone wires were buried everywhere as in cities.) Most people overlook these monopolies.

The only people paying close attention are the investors and speculators who want something for nothing, or, to be fair, a lot for very little. Being so remunerative, utilities appear as low-hanging fruit. Once plucked, the massive cash flow makes utilities into major political players.

Enclosures of Common Revenue

Not only can utilities force prices up, they can also force costs down. Of course, location has no cost of production (since nobody produced it and nature never charges rent). But beyond that, since money means power, once utilities have enriched themselves – grossing over $400 billion in 2017 – they pretty much have the public's servant (government) serving them.

Elected officials are supposed to be our faithful stewards, but often aren't.

- Utilities get to use public space (beneath streets) or common space (geosynchronous orbits) without paying the public much. Even better from a corporate standpoint, TV broadcast companies were able to win license to their channels without paying the Federal Communications Commission a penny.

- The state does not excel at keeping prices down. Public Utility Commissions routinely rubber-stamp every request from privately-owned electric utilities to raise rates. People who live in areas served by publicly-owned utilities pay much less.

- In Oregon, the major utility did not hand over to the state the taxes it collected. Rather than get fined or go to jail, management contributed to campaigns. The legislature passed another law requiring the utility to obey the first law. They ignored the second one, too.

Government fails to recover full market value for the fortune-making sites that utilities need. What kind of stewardship is that? Not exactly faithful (to the public welfare, that is).

The trend may be getting even less favorable to the public. Some localities are selling their water systems to private corporations. Water is right up there with oxygen as a life-or-death necessity. When you look at how many hundreds of times greater than cost that pharmaceutical companies charge for life-saving drugs, *one must wonder to what level will the price for water escalate?*

To avoid paying the public, businesses, naturally, prefer to turn public space and common space into their own private property (not that anyone needs privacy beneath a street or in orbit). Getting a never-expiring lease for some EM spectrum, which used to be entirely public property, for free, as TV networks get, was not good enough. Now, insiders and major corporations get to own frequencies, and the FCC lets these valuable gifts of nature go at well below market value.[1] Virtually, it's money for nothing.

The cloak of private property is perfect for claiming to not owe the public anything. Looking into the issue of the worth of rights-of-way, you

---

1   "The Failure of FCC Spectrum Auctions" by Gregory F. Rose & Mark Lloyd for *American Progress*, May 2006

find them referred to as the private assets of private corporations, belonging to them and them alone. Want to make something of it?

Anyone who uses a public or common space as their own will come to feel like it belongs to them. Restauranteurs who set tables and chairs on the public sidewalk feel like our public right-of-way belongs to them. Homeowners who park their car on the sidewalk feel the same. Whether encroachment by the less powerful among us or enclosure by the more powerful, it's not an unusual human behavior.

Even if commonplace, there's still the matter of scale. The homeowner and cafe owner might profit somewhat by encroaching, while the enclosing utility corporation will gain big-time. Getting free use of a location beneath a street or in the EM spectrum or in outer space does save the corporation immense sums. Those savings, you may have noticed, do not routinely get passed on to customers as lower prices.

Ironically, the issue of corporate ownership is irrelevant. Well, that is, if we set aside the power of money for a moment. In a practical sense, in order to draw revenue from natural monopolies, not only can government lease nature the public owns at full market value. It can also levy land and resources and spectrum that private parties own at their full, annual, monopoly rental values. Or try to. Utilities have been pretty good at not only dodging taxes but getting back refunds far bigger than any taxes they pay.

WORTH OF QUASI-LAND MONOPOLIES

Utilities, like landlords, land speculators, and mortgage lenders, enjoy being members of the tiny club of big rentiers. It's to their advantage that the majority are unaware of rent. It's to their benefit that conventional economists don't study it, that mainstream statisticians don't count it. Yet fortunately for us, somebody does, and their figures let us measure more of that stream of spending for natural goods and natural monopolies.

Were prevailing political winds to blow that way, and were government to charge utilities full-market value for the monopoly franchise it grants the company, how much revenue could the government raise? Consider the electric companies that deliver the juice. At a minimum, government could recover the difference between the profit of public and private utilities, about $33 billion annually.

That still leaves the value of the locations that utilities use. *In 2015, utilities (except for the ones the Census Bureau left out) paid local and state governments $167 billion,* presumably more now in 2019. Given the able resistance of the utility lobby to paying the public anything, those billions are likely to be well under the actual annual rental value of utility franchises and utility spaces. Closer to a fifth of a trillion is more realistic. It'd push our previous total (Ch 20) to $4.6 trillion. How much is that? Per capita, now the share would climb to over $2k per month for every registered voter.

MONOPOLIES FIGHT PROGRESS

While even 167 hundreds of billions are real money, how long will they stay real? Might techno-progress shrivel the value of a utility franchise? Or, exacerbate it? What will the advance of technology do to monopoly?

Could techno-progress make it possible to provide service without having to have a monopoly, letting in competition, and driving down price? It happened with telephones. If every home had its own power plant – say solar panels, as now it might have its own little garden – there'd be no need for electrical utilities. Already, a growing number of households are going off the grid. Not just "green" hippies but conventional companies, too, have turned to producing their own power.

It's a scenario not likely to warm the hearts of monopolists. And if we can foresee it, so can they. What steps are they taking? To date they've fought tooth and nail to maintain their advantage. Back when Edison and Tesla were competing to see whose electricity would dominate, the one that traveled long distances without losing lots of juice defeated the one more efficient at short distances. Yet today if DC were dominant, much energy and money could be saved. Some day, perhaps.

That was not the only battle. Look at the intersection of politics and technology throughout history. RCA, which founded NBC (and was granted part of the broadcast spectrum for free), kept television from the American public for 20 years, until the patents of the actual inventor expired, so to avoid paying him his just reward.

Corporate interference in the progress of technology is not a relic of the past. Just recently, utilities are making a push to revive nuclear power, which would keep consumers hooked to the power grid. And they're trying to keep solar from eating into their market share.

The only standing that corporate utilities have is a legal one, the government-granted title. They don't have moral tradition. Indeed, just as General cum President Eisenhower warned us about the military/industrial complex, so did Thomas Jefferson warn us about bankers and corporations. Utilities are just lucky that the tradition of the commons has faded enough to have become invisible to most modern humans. Otherwise, customers might demand an accounting – and probably lower prices, too.

Even if someday those utility monopoly rents do go away, right now today we have more sources of rents to find and measure. What else is natural and valuable, not just health-wise but also economically? Answer: all nature that's not owned but shared – in a word, the environment.

CHAPTER 23

Eco-librium Repatriates Rents

Somebody's out there. Sign seen in space: "Used planet, marked down, priced to go."

Myopic Grasshoppers Depress the Earth

Given the choice, would you rather live in pollution or without pollution? It's another way of asking if you'd rather save money upfront or save money downstream. Our choices do impact the worth of Earth in America, and indeed everywhere.

Ironically, while we cannot pay anyone to produce Earth, we routinely pay certain people who, to some degree, destroy Earth. We pay owners of:

- utilities whose power plants pollute the air basin;
- companies that develop housing tracts sprawling over cornfields;
- logging corporations that clear-cut mountain sides which ...
- silts up the stream you once fished;
- maybe even causes a landslide that dislodges your house;
- mining companies whose tailings contaminate groundwater;
- chemical firms whose windblown GMO pollens contaminate farmers' fertile soil; and
- agri-businesses who treat soil like dirt, decreasing its fertility;

If you choose to save somewhat upfront, yours is a world of opportunity. It's cheaper living in the shadow of a skyscraper that blocks out sunlight than in the penthouse with a view of 360 degrees; beside a freeway with constant noise and smog versus even a block away with cleaner air and a measure of quiet; and near a landfill perfuming the neighborhood versus far away.

If you can not afford environmental health, likely you live where land values are already low. Then add a trash incinerator (or prison, weapons factory, or tobacco subsidy). Land values won't always fall and may even rise, since tax-supported jobs pay better and developers respond with better houses.

While these incomplete lists tilt toward producers, consumers too choose convenience over long-term health. Most of us drive. Some home-owners spray poison on their lawn. Truckers leave their diesel engines running. Shoppers prefer packages of "food" (low nutrition, high additives). Households contribute generously to landfills. It's an attitude.

All this abuse makes land sad and depressed. Our methods of meeting our wants and needs chew Earth at one end of the economy and spew toxins into the environment at the other. Going about our business that way, harming the health of Earth, we depress the worth of Earth, *ceteris paribus* (all else being equal).

While people pay less for despoiled land upfront, the despoliation forces people to pay much more for other things downstream: restoring the environment, healing ourselves, and for inefficient production. Were we not despoiling Earth, we could use those payments to build up her worth. Hence those payments for rehabilitation are surrogate rents and count toward a complete total for the value of land and resources.

"Leaders" Limit Liability

How did we ever become such a careless, messy civilization? While polluting and depleting are business as usual now, they were not always quite so usual. Hence industry turned to the legislature.

It started out benign enough. Long ago, it used to be that if a neighbor erected a tall building that blocked your sunlight, you could appeal to the king or local noble. And win.

So when a village wanted to do something collectively – say, build a watermill – the group of residents who actually made it happen became a temporary part of the local government. That is, they *incorporated* into the body politic ("corpus" means "body") and enjoyed the council's limited liability. When the project was completed, the corporate charter and the limited liability with it expired.

But investors could not leave well enough alone. Thomas Jefferson, a Founding Father very interested in the power in the land and with an insightful opinion about many topics, lamented that the younger generation, lacking the *"...principles of '76 now look to a single and splendid government of an Aristocracy, founded on banking institutions and monied corporations...."*

Then came James Watt's steam engine and industrialization. If a train locomotive roaring by set your cornfield afire, you could sue – and actually win. To avoid restitution, or even bankruptcy, for businessmen limited liability was too good an idea to ignore. Exerting the pressure that comes with impressive profits, industrialists persuaded politicians to grant them a corporate charter. What once covered temporary nonprofit groups acting for the commonweal, now limited the liability of permanent companies acting in their own self-interest.

To limit their liablity personally, the businessmen – "corporations" in the US, "limiteds" in the UK – paid no more than a small filing fee. If found liable, management fobbed off the fine or penalty from themselves, the responsible humans, onto an abstract corporation; those who ultimately paid were stockholders and consumers – even taxpayers. The real perpetrators in management skated.

In the US, individuals faded into the background and corporations became "persons." A railroad lawyer who was a clerk for the Supreme Court wrote in his own hand on the cover of an unrelated ruling that corporations were persons. The Justices and everyone else even up until today accepted his radical penmanship.

PURCHASE LIMITED LIABILITY

The problem with pollution is like the problem with politics. In politics, those few who band together to gain a lot have much more success lobbying than those many who stay disorganized and individually lose a little. Similarly, those few who pollute a lot do so with impunity while those many who receive a little poison seldom defend themselves – and rarely do they win.

For a business that is a bad neighbor, the more damage you do, the more you need a limit on your liability. For those guys, after the land title, the corporate charter might be the second most valuable piece of paper dished out by government. Heck, only a direct handout or bailout – cold cash in the pocket – could be better.

Trying to cope with corporations, well-meaning critics of money in politics rail against the more recent ruling, *Citizens United.* But they're barking up the wrong tree. There's a simpler, more fundamental solution.

Charters need not be quasi-freebies from the state. Government could charge full market value for granting them. Or government could get out of the liability business altogether and let insurance companies (they contribute the "I" in "F.I.R.E." along with Finance and Real Estate) handle it.

To stay afloat, insurance companies would not be so accommodating but charge full market value for limiting liability. Businesses imposing risks on others – polluters, depleters, and wasters – would pay premiums of staggering amounts. Firms not endangering consumer, worker, or nature would pay less. Until that fine day, we'll still have more lawyer jokes than economist jokes.

Just as economists don't discuss land much, most don't discuss environment, either, except to refer to its debasement as an "externality." The term implies that such damages were unavoidable consequences with no one at fault, instead of results of irresponsible behavior, like littering or reckless driving. Thus academics defuse the responsibility of wealthy and powerful industry. If your tracks need covering, it must be nice to have an entire discipline to do it for you. Yet can an apologist be a scientist?

Conversely, economists don't refer to the value of land as a "positive externality." Yet the value of a property is not due to the owner who gets to sell it or lease it. Rather, as everyone knows, it's due to location, location, location – to natural features like a good view and social features like a nearby transit stop. By not referring to land value as an externality, economists deprive society of its authorship and look the other way when owners reap something for nothing. As a positive externality, publicly recovered rent would not take any private value from anyone.

Farsighted Ants Enrich the Earth

While we think we save money by being sloppy and not cleaning up after ourselves, we don't. It's not just that we raise costs for society, for people downstream and downwind, we also miss out on cutting our own costs. Wasting and leaving behind waste is wasteful, a loss. A win/win is a company that captures sulfur from smokestack exhaust; it thereby has another product to sell. Efficiency is a savings, an untapped resource. An engine that burns all its fuel not only leaves zero residue to pollute but also gets more miles per gallon.

For some, that's too good to be true. Regardless, other humans are evolving. Many have reached the point where they regard pollution as much a social *faux pas* as littering, smoking in public, or spraying DDT. They might raise the bar.

In response to higher standards enforcing our right to a healthy planet, some businesses would raise their prices – and lose customers, market share, stockholders, share value – possibly go broke. Other businesses would make money by being efficient. Smart dirty industries would convert to clean technologies – which in most cases already await them on shelves. Clean companies would gain market share.

With bigger operations, clean companies could streamline, enjoy economies of scale. Both the always clean and the newly clean would see their cost of doing business fall. Already firms that win "green" certification don't lose money but save. With the extra wherewithal, these successful businesses could spend more on a better commercial location.

Customers, too, would have more money to spend on non-polluted land. And they could do so, even if their incomes did not rise, because their cost of living would fall. First, due to businesses competing among themselves, efficient firms would lower the prices they charge their customers.

Second, as the environment heals, so would humans, enjoying air, water, and food without added toxins.

Humans would spend less on doctors and medicines. And as usual, due to competing among ourselves for desirable locations, we'd spend more on land, pushing up its value.

Thus, one can estimate the missing value of the environment from its balance sheet: (1) downstream payments that degradation imposes on us, such as dealing with radioactive waste; (2) the loss in value of degraded areas, like neighborhoods near toxic dumpsites; (3) a portion of spending on medical treatment; and (4) the savings from greater efficiency in doing business. Combining the four tells us how much more we would have available to bid up the rent for living and working on healthy land. A hefty portion of that is by how much the worth of Earth in America would rise.

COSTS OF DAMAGING LAND

For up-to-date figures, let's google combinations of "limited liability" and "corporate charters" and "corporate damages" and "negative externalities" and "environmental harm" and "environmental illnesses" and "pollution" and "imposed costs" and "true cost" and "total costs" and "percentage of profits due to limited liability."

If a business did not harm a customer, worker, or nature and get sued and pay lawyers, it'd have lots more money to invest elsewhere. In 2008, eLawForum estimated litigation cost corporations $210 billion, equivalent to one-third of the after-tax profit of the *Fortune* 500. Not all those suits were filed because of pollution but some were. On the other hand, not everybody harmed by pollution filed suit.

Researchers base calculations of environmental costs in part on court fines mulcted upon degraders. Those values exist on paper, but not so much in the pockets of plaintiffs. Collecting the fines is problematic.

- Residents of places like Louisiana and Alaska have yet to receive what oil companies were supposed to pay for ruining their ecosystem.

- Wall Street financiers were supposed to pay victims and local governments for fraudulent lending to buyers of homes+homesites.

- Tobacco and pill companies did pay their consumers something but less than what they were supposed to pay.

It's a pattern. Where's a government when you need one to enforce a fair ruling?

In estimating how much the public loses, most researchers don't present separate totals but subtotals or grand totals. Combined into a mosaic, it paints a pretty pricey picture for environmental damage.

Not in Mexico but in America, drinking tap water sickens 20 million annually at some cost. Treating fresh water, polluted by agri-biz's phosphorus and nitrogen, a decade ago was costing over $4.3 billion annually. Cleaning up the 126,000 polluted groundwater sites in America would cost from $110 billion to $127 billion.

The true cost of emissions, factoring in costs of premature death, illness, increased loads on the healthcare system, lowered crop yields, missed work days, higher insurance damages from extreme weather events linked to climate change, etc., pumps up the price of a gallon of gasoline; in 2015 you'd pay $3.80 more at the pump.

Costs per kilowatt hour of generating electricity is 10 cents for coal, 7 cents for natural gas, 13 cents for solar, and 8 for wind. However, when you add in environmental and health damages, costs rise to 17 cents per kilowatt hour for natural gas and 42 cents for coal. Other researchers found:

- Air pollution caused by energy production in 2011 alone cost at least $131 billion;

- Coal, from mining to burning, costs the public $175 billion to $500 billion annually;

- In 2015, damages from producing electricity in the US totaled $330–970 billion a year.

In 2008, children sickened by our altered environment (smog plus toxins) cost adults $76.6 billion, equal to 3.5% of total illness costs. That figure leaves out sickened adults. Obama's OMB in 2013 said *carbon damage alone is costing us $200 billion.*

CARS & THE CULMINATION OF COSTS

Lung disease caused by breathing air-borne pollutants, the American Lung Association said in 2014 cost at least $130 billion annually.

Probably the most polluting thing individuals do is drive. Our vehicles' pollutants in the air we breathe exacerbate asthma, shortening the lives of newborns and the elderly. Furthermore, the incidence of autism and Alzheimer's is higher along high-traffic streets. But because those roads are busy, which raises location value, it's hard to tell how much smog lowers site value.

We tolerate polluted air in part because driving is convenient – or so we think. Ivan Illich added the time it takes to make the money to afford driving and added it to time sitting in our cars – crawling around sprawl far from the lively city center or stuck in traffic, feeling frustrated while worsening our well-being. He found cars actually carry us at four miles an hour. Of course, very few make that calculation, since they don't have to drop a quarter into a meter every quarter mile for fuel, repairs, insurance, purchase, fees, smog damage, health impairment (cars are fattening), etc.

Failing to charge drivers the equivalent of rent for using roadway land underprices driving. That zero charge thus underprices roads compared with other uses of land. Hence society devotes extra land to roads – a major source of pollution from engines and tires, and less to walking paths, bikeways, light-railways, and other modes that pollute not at all or much less. Making roads appear free creates more automobile dependency, which encourages more sprawl onto farmland.

In 2014, Eosta, which focused on waste in agriculture, put global costs of eco-damage at $4.8 trillion. A year before, TruCost added agri-business (not normally publicly traded and so left out before) to their list of culprits damaging the planet. They upped their estimate of a global cost to $7.3 trillion, meaning the US share would be around a trillion and a half.[1] Ralph Estes in his book *Tyranny of the Bottom Line* factored in workplace injuries, medical care required by the failure of unsafe products, health costs from pollution, and many others. He figured that external costs to US taxpayers totaled $3.5 trillion in 1995 – four times higher than the profits of US corporations that year ($822 billion).[2] Nearly a quarter century later, it's likely higher.

LOSE LOSSES, GAIN GREATER RENTS

Just as the estimates for total land value were not easy to come by and were all over the place, it's the same with the estimates of land dam-

---

1   "Natural Capital at Risk: Top 100 Externalities of Business" at TruCost, April 2013
2   "When Good Corporations Go Bad" by Erik Assadourian in WorldWatch, May/June 2005

age. Another caveat: Once humans stop fouling their nest, they'd not use every saved dollar on bidding up the value of the location they want for themselves. But some of those saved dollars would go to that purpose.

Using TruCost's $1.5 trillion as a floor and Estes's $3.5 trillion as a ceiling, let's go with the midway $2.5t. If Americans got to save all that, they might spend $1.4t on bidding up the value of land. Adding the cost of degradation to our growing total ($4.6t, Ch 22) would push the overall total for rents in America to about $6 trillion (how much BlackRock manages, Ch 12). Per capita/registered voter that is $3k/mnth.

While polluters and depleters might worry about society making degraders pay, why worry? Once we heal Earth, we not only heal ourselves but we also enrich whoever ends up with the fatter rents for healed land. Polluters and depleters can shift their portfolios and invest in real estate.

Along with corporate charters, waivers of standards, and lenient enforcement, government grants other little pieces of paper worth trillions. These new official forms likewise grant monopolies and are major sources of rent. The streams of spending they corral also need to be added to the total flow of what we spend on the nature we use.

For us to tabulate the worth of Earth in America in total, we need to know more than the value of land and resources, more than the value of the unowned environment, and more than the padded profit of land-based monopolies like utilities. We also need to know the padded profit of monopolies based on the very land-like field of knowledge. That is, we need to know the rent from holding patents and copyrights, whose purpose is to exclude competitors.

CHAPTER 24

STAKING CLAIMS ON FIELD OF KNOWLEDGE

"Despite being fact-filled, this article is more than 99.99% empty space."

NO TRESPASSING ON FIELDS OF KNOWLEDGE

Humans of our era understand that matter is an expression of energy, energy is an expression of subatomic particles, and subatomic particles are an expression of natural laws. Our era is coming to understand, but is not quite there yet, that matter, energy, and the laws of physics are all part of nature. The realm of logical solutions is as much a part of nature as is planet Earth. All of it exists without the input of any human's labor or capital. And the best parts – the best locations – are extremely valuable.

Cities can expand, but there's only one downtown location at Broadway and Main. Knowledge expands but there's only one algorithm for the fastest web search. Whoever gets to own the best locations or the most useful knowledge gets the chance to rake in the most money.

Besides applying "hers" labor and capital, that owner could exclude others from the best that nature has to offer. Minus competitors, that owner can grab an extra, unearned profit. And besides keep out others – as absentee ownership is the granddaddy of all privileges – that owner can step aside and let in others to use "hers" land or idea, for a price, like a troll under the bridge permitting passage. Both that price and that extra profit are rent. Along with rent for environment and utilities, those incomes augment the worth of Earth in America.

To spur ingenuity during the dawn of the Industrial Revolution, the authors of the US Constitution included patents and copyrights. P&C, like land titles, professional licenses (e.g., medical), corporate charters, and utility franchises, also took nobody's labor or capital to create – other than the labor to lobby and the capital to make campaign contributions. As Americans spread from the East Coast across the continent, they won from the federal government land patents granting a monopoly, over a sector of land instead of over a section of harmonious possibility.

Getting a patent or copyright is like planting a flag somewhere on the realm of logical solutions. If the location is a good one, by posting that "No Trespassing" sign the claimant can exclude others and charge a rent. Since government does not charge full market value for its patents and copyrights, the amount that the government does *not* charge is also rent.

FROM EXCLUDING TO DEPRIVING

Being able to exclude others from land or the realm of harmonious possibility can get out of hand. While most of us make money by doing something useful, like being a doctor, a few get paid by preventing others from doing something useful. That's anti- social.

> • Centuries ago big landowners, backed by the power of the state, got bigger by enclosing common land, preventing "commoners" from using it. Only the big owners could farm or graze sheep on land that once was available to all.

> • In our era, the AMA prevents a doctor from France – where the lifespan is longer than in America – from practicing medicine in America. That way, they decrease competition, swell their market share, overcharge their customers, and rake in more money.

This rent differs from that of utilities. The doctors' monopoly is not natural but artificial. While it would be impractical to have parallel sewers compete, it'd not be impractical to let all doctors compete. On the contrary, it'd be quite efficient to have more doctors competing among themselves. And with experienced nurses, too.

Owning aspects of nature differs from owning things that we do create. When you keep others from using your house or car or computer, there are plenty of other houses or cars or computers they can turn to, since people make them. But when you speculate in land, and your vacant lot keeps others from using a prime location downtown, then you hobble your local economy. Similarly, when you keep your idea to yourself and prevent others from building on it, then you hobble scientific progress.

DISCOVERY

Before any humans discovered North America thousands of years ago – whether entering from Siberia or Scandinavia – it existed. Before anyone discovered waves of gravity, they existed. And before anyone worked out the functions of trigonometry, they existed as logical patterns, the only forms that could exist. Long before humans were able to discover them, such useful parts of nature were already available. Then each discovery underpinned the next.

Just as long voyages to explore virgin territory depended on already occupying a home base and employing logic, so did learning the ways of the physical world depend on a pre-existing base of knowledge and following reason. As Sir Isaac Newton said, "If I have seen further than others it is because I have stood on the shoulders of giants." Those shoulders belonged to thinkers like Copernicus and Bruno (who was burned at the stake). Later, if *James Clerk Maxwell had not died young, he may have discovered e=mc2* before Einstein but at least he paved the way for Albert.

Some discoveries don't pay off in money. Indians were first to populate the Americas but did not make much off real estate speculation. Teams of scientists collaborated to discover gravity waves but none became rich in the process. And neither Isaac Newton nor Gottfried Wilhelm Leibniz marketed calculus. Only those who came after profited from such discoveries, usually due to excluding others.

Our privatizing of knowledge goes back millennia. The ancient Phoenicians gave their captains the order to sink their ship before allowing a pirate or foreign nation to capture it and discover the secret that allowed Phoenicians to sail from one end of the Mediterranean to the other – only the Punic sailors knew how to caulk their ships with tar. Centuries later, other Arabs kept secret the formula for steel which made their scimitars so much more deadly than heavy iron swords.

Ownership of originality is huge in this modern America of constant progress and incessant litigation. For its part, technology has made major strides rapidly. The rather mundane doorknob is actually a rather recent invention, patented in 1878 (a year before land reformer Henry George's classic, *Progress and Poverty*).

DIBS

Although two people cannot occupy the same space at the same time, two people can utilize the same equation at the same time. My acquiring knowledge does not mean you have any less knowledge. Yet discoverers of logical solutions deserve reward. Do we moderns owe the descendants of Newton and Leibniz untold fortunes? Similar to how much we pay today's authors of code, such as the Google guys?

How far do the benefits of being first extend? Could the first crew of canoeists landing on what's now Alaska claim all North and South America? Could the crew member who was first to set foot on the beach be the one to be entitled to the Western Hemisphere? And for how long? Forever? By now, with every inhabitable corner of the planet already inhabited, nobody can claim to be first. Nobody can even claim to descend from whoever arrived first, since tribes have always wandered all over the face of the earth.

Being first confers the right to keep the discovery and exclude others from it. Even children know to be first to call "shotgun" and claim the front passenger seat. "First come, first served."

But what if two people discovered an aspect of nature at the same time? The history of science is filled with smart guys breaking through at almost identical times. Nearly simultaneously:

- Newton and Leibniz calculated calculus;
- Lavoisier and Priestly isolated oxygen;
- Darwin and Wallace theorized evolution; and
- Fritz Hasenöhrl, precursor and contemporary of Einstein, reasoned his way to a more complicated e=mc^2.

It's like something was in the air, that a few antenna could pick up. Should these pairs split the benefits 50/50?

The one to win a patent for the telephone, Alexander Graham Bell, walked out the door of the Patent Office as another inventor of the telephone (Elisha Gray) walked in. Should the guy a few minutes late win nothing and the guy a few minutes early win everything? What if the person who patented or copyrighted first was not the one who discovered or invented first? Snooze, you lose? Everything forever?

Paper Claim

Patents and copyrights prevent late-comers and anyone else from exploring that particular part of the physical world and deducing a similar correct answer. People accept the notion that P&C protect the little guy – the basement inventor, the unheralded author. Yet it's a star system; only a very few inventors and authors make any money from their patent, or composers from their copyright.

While patents and copyrights are justified as ways to encourage creativity, they do just the opposite. Big companies like IBM get literally thousands of patents each year, and little companies called "trolls" get bunches of patents, too. The bigs and not-so-bigs do this not to use the aspect of nature that they've staked out, but so that you can't. They're a dog in the manger, putting a roadblock in the path of progress.

With government-granted patents in hand plus one more major favor, a favored firm can preclude competition, dominate the market, and achieve near monopoly status.

- Ford for a while was untouchable thanks to patents plus gangsters purchasing getaway cars and police forces, too, to chase them;
- Johnson & Johnson is still the biggest pharmaceutical due to patents and limited liability (drugs do harm, too);
- More recently, giants Amazon, Apple, and Facebook benefited from patents and government-funded research.

At its core, a successful business has a great new idea. But a giant business at its core has that plus a stockpile of patents. Giantism cannot flourish in a truly free market. It can only come about with the helping hand of the state. Government assistance – that's where super wealthy families come from and what turns I.T. multi-millionaires into I.T. multi-billionaires.

Ironically, today's tech giants may be on their way to utility status, or even there already, and become subject to regulation. Society may legitimately wonder, with this right to exclude, does any responsibility or duty come with it? Like, you may own it, but for excluding every other person from that part of nature, do you owe them compensation for never letting them go there? Do humans have an equal right to all parts of nature?

GO INTO BUSINESS

A more mature strategy than shouting "dibs" in order to be first is taking turns – as harried parents try to teach their brood. For brainiacs and artists to take turns exploring the realm of logical possibility, patents and copyrights would have to expire much more quickly than now, after a couple years, not decades.

However, bowing to investor pressure, US politicians have lengthened P&C; patents were for 17 years, now it's 20. That's just the reverse of the adage, "the penalty does not fit the crime." Here, the reward does not fit the creativity.

Sometimes a discoverer can get a patent but never make a penny. Usually the gizmo is not marketable, but often the innovator gets ripped off. Almost all inventors and artists are too poor and desperate to have the leverage to negotiate a fair contract. Hence there are innumerable cases wherewhere recording artists make peanuts while record companies keep hundreds of millions from the sales, and of people, like the inventor of the windshield-wiper, fighting Ford his entire working career to win the profit he was owed (he eventually won the case, but lost his family in doing so).

Most creative types are not the best business people and need a partner. Paul Allen needed Bill Gates and Steve Wozniak needed Steve Jobs. Henry Ford nearly wrecked his company until he listened to those with business sense and stopped making every car black.

Sometimes, a discoverer can eschew the patent and make a bundle. Usually with a partner, the team take the new idea to market before anyone else, become the best known to consumers, and get a head start on any competition. They "corner the market," make a pile of money, then even more by maintaining the largest market share. All that gain would be due to their labor and capital, none of it would be due to keeping everyone else out of that particular arena. None would be rent.

FEES

For granting a patent to an unsellable invention, government charges the same amount as for a patent protecting a hot new app. For granting a title to a quarter acre in Death Valley, government charges the same amount as for a title to a corner lot in midtown Manhattan. In cases of the app and Manhattan, the gulf between filing fee and remuneration yawns like the Grand Canyon. For next to nothing, holders of patents and deeds get to exclude everyone, for a long period of time, and win this without having to compensate society.

If it were a business that issued patents, how much would they charge? They'd not charge everyone the same, but as much as the patent were worth. Pressured by such logic, the government did increase its application fees recently, but not much.

Would inventors still invent if they had to pay full value for a patent? Would writers still write? Of course. People with ideas love to see their ideas in the world. That's what drives them, not a piece of paper, not a monopoly.

Some creative types even forgo patents and profits. Look at Linux and the rest of Freeware in our current economy of code. As Jonas Salk (and later Ralph Nader) said, how can you patent the sun?

In the mid 1950s, when polio was crippling kids (earlier President FDR), Jonas Salk and Albert Sabin developed an effective vaccine at nearly the same time. Both chose not to patent their vaccines worth millions, maybe billions. By the way, outside the US, Sabin is more famous for eradicating polio than Salk. And both built on the work of the unheralded Hilary Koprowski.

While those researchers may have had feeble profit motives, some businesses with a robust profit motive declined to patent and still struck it rich. Coca-Cola is the classic. Oracle and Google, both worth multi-billion-dollars, had almost no patents until a few years ago. Then, notes law professor Dennis Crouch (private email), they won patents not to establish dominance but to solidify dominance.

Rather than hinder progress, a patent fee in line with actual market value would spur inventors and investors to form partnerships. Same with filmmakers and distributors. Nobody would sit on a good idea. Just the opposite of now when an owner of a prime site can let it lie fallow, waiting to cash in later.

Cheaper than patents for discoveries are copyrights for creations – they're free. Because creations such as stories (some say there are only three ... or six ... or nine basic plots) can be retold endlessly, even if copyrighted, society loses nothing. Often a creator does not want a copyright. When you hear a new joke, you have no idea who first told it. And why should a jokester not freely contribute to society?

P&C COMMAND DOLLARS

It's standard practice for the government to let Google, Microsoft, Yahoo, and many other tech companies use copyrighted material without a license from the copyright holder. In 2007, the Computer and Communications Industry Association (which includes the tech giants) figured this legal evasion accounted for much of the growth of the previous decade and generated more than $4.5 trillion in annual revenue.[1] How much of that would tech companies be willing to pay as rent? A third? $1.5 trillion?

McKinsey & Company, which gets coverage in the *Wall Street Journal* (e.g., Jan 10, 2007) for tracking financial assets worldwide (totaling $140 trillion in 2005), estimates that as much as 80% of stock price is accounted for by patents and copyrights. The US stock market in 2018 is $30 trillion, so IP is $24t. Converting from price to rent puts rent at $3t.

Kevin A. Hassett and Robert J. Shapiro, who do estimates for a living, calculated a total for intellectual capital (patents, copyrights, databases, and general business methods). In 2011 it was as high as $9.2 trillion. Since that figure approaches the then-current GDP, those guys must have meant price, not annual value. Rent would've been $.92t.

In 2013, a partnership of statisticians for private business put patents at over $5.8 trillion, almost half of the then GDP. The US Patent Office also estimates how much IP – including IT workers and sales and the

---

1   "Fair Use Worth More to Economy Than Copyright" by Thomas Claburn at *Information Week*, 12 Sept 2007

leverage of exclusion – contributes to the economy.[2] In 2016, they figured over $6 trillion,[8] close to the $5.8t three years earlier. Since neither figure is a lump sum price but an annual flow, how much of that is rent – profit due just to exclusion? It's hard to say but $2 trillion is reasonable.

Add that $2 trillion to the $6t we reached by adding utility franchises and eco-losses to land and resources; the new total reaches $8 trillion. That's equivalent to the drop in global stock prices in January 2016, to what it would it take to wipe out the US federal deficit, and to how much builders worldwide spend in one year on putting locations to better use. Per capita of registered voters comes to $4k per month.

Eight trillion approaches half of total national income. That's bigger than either the returns to labor (mostly wages) or to capital (to lenders and investors). (With land, labor and capital are the other two factors in production.) That means the value of our output is due less to anyone's input – labor or capital – and more to nature and privilege. In other words, you spend more for what nobody created – directly and indirectly – than you pay others for their efforts.

If $8t sounds unbelievable, it's going to get worse. We've reached this height without yet figuring in two more major sources of rents: finance and taxation. Because rent is much money for doing nothing, it attracts speculators. So we go where the money is, as Willie Sutton said, with those too big to fail. Me burning up my adding machine blazes a new trail.

---

2   "Intellectual Property and the U.S. Economy" at USPTO

CHAPTER 25

SCOUTING AHEAD OF THE CURVE STRAINS TIES

"Out on a limb, above a well-guarded secret, only your balance has got your back."

WORKING WITHOUT A NET

I t's not like economists and statisticians to ignore big numbers; normally, they don't. But talk about the elephant in the room; $8 trillion for the worth of Earth in America *annually* (Ch 24) definitely qualifies as huge. So, why'd they let us go out on a limb by ourselves?

Only a few researchers make educated guesses and those always slight the land. Even at eight trillion dollars, the count could be an underestimate. Under-estimate is what officials do. Getting clear is what they don't do (Ch 18). Nor do economists, and bureaucrats after them, show interest in noting the two kinds of spending. One rewards the efforts of workers, the other rewards the ownership and privileges of others.

Allies are not found among critics and would-be reformers of standard economics. They don't criticize the inadequate treatment of land nor join the call for calculating rent. Ironically, the issues they find relevant exist only because they're downstream of the rent problem.

However, our real opponents are found among the rich and powerful who now receive the lion's share of our spending for nature and privilege. Lenders of mortgages, owners of oil companies, holders of patents, exercise their considerable power to keep real rent statistics under wraps (Ch 12). Even the middle-class, with inconsiderable power, prefer to ignore that their lesser captures of rent are unearned.

We questers of the size of rent are pretty much on our own, tallying staggering figures, walking the high-wire alone, while working without a net. If we get the total wrong, that means ignominy for this quest – by us or anyone to follow us – for a long time to come.

Most cannot see the land, but only what's on it – unaffordable housing.

Urban advocates miss:

- What rises in value is not housing – already built – it's land. Further ...

- Sellers and landlords don't cause the rising; it's residents with more money to spend on prime locations.

Many criticize the GDP as a measure of economic health and a guide for policy and offer ideas to reform it. Yet, few notice that the strongest stream in the GDP is our spending for parts of nature. People worried about GDP, housing, and inequality, leave us to our own devices.

ACADEMICS CRITICIZE BOGEYMAN

Thomas Piketty became a celebrity with his fat bestseller, *Capital in the 21st Century* (2014). He showed how the rich get richer while the poor get poorer due to ownership (or lack) of capital, or so he said. Piketty used "capitalists" and "the rich" as synonyms and proposed taxing them. Similarly, the tax code calls income derived without having to apply labor or capital "capital gains", rather than "land gains" or "privilege gains".

Reed College (in Portland Oregon) graduate Matthew Rognlie while still a grad student at MIT noted that capital—factories and skyscrapers—depreciates; it's land—especially urban land—that appreciates. As do most in his discipline, Matt mislabeled land as "real estate". What captures the surplus that functional economies naturally exude is not buildings or any sort of capital but land, becoming the father of enduring fortunes.[1]

Sitting side by side with Piketty on a televised panel discussion, Joseph Stiglitz echoed Rognlie, saying it's not owning capital that widens the income gap but owning land. Piketty did not acknowledge the point. And making the point was the former Chief Economist of the World Bank, a winner of the ersatz "Nobel" prize, and a professor at Columbia.

How many know Piketty's idea? Millions. How few read the correction? Several thousand? Even when getting it wrong, a critic of wealth inequality can achieve fame and fortune, leaving geonomists as lone voices howling in the wilderness.

Certainly, a few capitalists did rake in fortunes. Railroads and steel, oil and cars, did vastly enrich a few. Yet the textbooks leave out these facts:

- Railroads made more money selling the land they were given by Congress than by ferrying freight.

---

1   "Meet the 26-year-old who's taking on Thomas Piketty's ominous warnings about inequality" by Jim Tankersley, Economic Policy Correspondent, *Washington Post,* 19 March 2015

- Iron, the basic ingredient of steel, was already valuable in the ground, untouched by miners (that value is rent).

- Oil, like iron, was already valuable even before being extracted from the ground.

- Cars needed paved roads; funding them came out of the public purse. Plus, courts permitted the destruction of the trolley system, which reduced competition early-on in the second industrial revolution.

Today, tech giants come from creativity and government-granted patents and subsidies (last chapter). None but geonomists show how rent and privilege played a huge role in amassing monumental fortunes.

CRITICS OF FINANCE OVERLOOK FUNDS

In keeping with their penchant for changing names and definitions, economists no longer use "capital" to refer to heads of cattle (the term's original meaning) or to tools, factories, and supplies (in contrast with consumable goods). Rather, they just mean big piles of money, stuff of investments and huge savings accounts. An enormous portion of this capital is curdled rent, invisible to critics and reformers – steroids for bankers.

Gerald Epstein and Juan Antonio Montecino, in their *Overcharged: The High Cost of High Finance* (2016), measured how much bankers pad their profits. Bankers' higher charges cost borrowers more money, which in turn harms the economy.

They use "cost" in the sense of "harm" or "damage," not in the sense of an input's expense. The two senses differ greatly; economics would benefit by keeping them straight. (So, when did consistency become a virtue in economic thought?)

Epstein and Montecino find that:

- Starting in 1980, investment and commercial banking's share of intermediation (middlemen shuffling money) began rising sharply, from about 20% to almost 70% during the first half of the previous decade.

- The IRA 401K pension system is bloated and non-competitive. It creates hundreds of billions of dollars in annual costs (asset management and advisory fees). Yet it does not create better allocation of resources.

- The share of financial sector "earnings" relative to total corporate profits rose from about 10% around 1980 to about 40% at the turn of the 21 Century.

- Despite their best efforts, the net returns of actively managed portfolios are more than 2% points lower than net returns relative to index funds – 4.73% vs. 6.94%.

- In general, the asset management sector generates significant amounts of income for themselves, less so for their customers and society.

- Beginning in the 1990s, the gap between wages in financial services and other sectors started increasing, and that gap was especially high within investment banks and securities brokers and dealers.

Building on Philippon and Reshef's research, Epstein and Montecino estimate wage rent (excess wages with excess profits) to be $1.4 trillion between 1990 and 2005. Adding losses from slower growth and recession brings it up to $3.7 trillion. They project that from 1990 to 2023, this number would add up to $22.7 trillion.

Should we add wage rent to the growing total of natural rent? What would be the annual amount of all sectors in 2017? I could not find statistics, so whether it belongs with natural rents or not is moot.

Leading up to the recent recession, the American financial system (by which Epstein and Montecino mean Wall Street, not credit unions and community currencies), increasingly failed at providing basic banking services and became more involved in ground trafficking. Their solution? Rules and regulators, no matter how much the bureaucracy would cost.

REGULATION OR AN OUNCE OF PREVENTION?

People love to hate the government – it's bloated, it's wasteful, it's all about favoritism, etc., etc. – yet we want more of it. Like Woody Allen's joke about the two Jewish ladies complaining about the terrible food at the resort they went to in the Catskills. Not only bad – there was never enough of it!

Most claim that not enough government – de-regulation – caused the recent recession, and more government – re-regulation – will prevent the next one. But they miss that:

- Government loosens the reins on financiers every turn of the business cycle.

- More critically, the life's blood of bankers is debt for land.

For a thorough history of finance, see Phil Anderson's *Secret Life of Real Estate and Banking* (2008). All that attention paid to what bankers do, and none paid to what they do it with. Overlooking rent legitimizes surface issues and marginalizes a deeper analysis.

Engineers say that if a problem can be defined, it can be solved; a good prognosis depends on a good diagnosis. Conversely, operating on the basis of a misunderstood problem, one can never conjure an effective solution (except by luck).

The role of the state as the main cause (or cure) of collapse pales beside the role of rents, our payments for the essentials of life. Only after we've acquired food (from land) and housing (on land) do we then purchase furniture, appliances, cars, etc. When we must pay more for the former, we can only purchase less of the latter (a recession).

The world's foremost organic gardener offered advice. Author of *The One Straw Revolution,* Masanobu Fukuoka said, "When something goes wrong in my garden, I think first not what to do but what to undo." Applied to financial collapses, regulations and the rest of the apparatus would be what not to do. But what would be what to do? Find another use for rent, other than reward speculation?

Howl Until Heard

What could show others the error of their ways? The harder you try to intrigue the likeliest receptors, the more stubborn they get. Your facts not only fall on deaf ears but also raise pointy hackles. Psychological research shows that people set in their ways merely entrench themselves even deeper into what they "know to be true." Staying loyal to ideology suits human nature.

Do the gatekeepers – mainstream economists and tasked bureaucrats – see this investigative gadfly as an amateur, an agitator trying to rock the boat? Does merely asking a question seem like prying? Imply criticism? Appear to be taking sides?

Once you know the role of rent and the magnitude of rent, the only thing left to do is be a lone wolf howling in the wilderness where the trees are silent, to howl with each rising moon and each Earth rise, until the howl is heard. Totals for the surplus output of the economy must get out – as must the story of why the effort is so lonely and difficult.

Could the number for the size of rent be found stashed somewhere? Could the total of the value of land, resources, and privilege be greater than already calculated? Could a smoking gun – a memo calling for the proper ensconcing of rent – be found anywhere? If so, where? Where is a sump for all the significant statistics? And who towers over the myriad of number crunchers?

As they say, follow the money. Who makes the most money off the Earth? What institutions do those aristocrats control? Let us soldier on into the belly of the beast, the Federal Reserve. To whom else can we turn? Can we get some ten-year-old kid to hack the Fed? (Joke). Will finding out the size of our spending for nature and privilege – our social surplus – wake up Leviathan?

CHAPTER 26

THE FEDERAL RESERVE – THE VAULT OF THE STAT?

"The Fed is the Washington branch of the New York banking establishment."

IN THE BELLY OF THE FED

I f anyone knows the worth of Earth in America, a good candidate is the lender to lenders, the bank of all banks, the US Federal Reserve. The Fed has the means and the motive to know the value of land and resources.

- It has the power to tabulate all rents, what with its huge staff and ability to demand data from member banks. And …

- It has a good reason to know the size of that spending stream. Knowing how much society spends to use locations, the central bank can predict how big its income will be, and when a current wave of rent may recede.

While the Fed never has to worry about a recession hurting it directly, it can still cash in by granting favors that must be paid back later.

What does the Federal Reserve know about the size of the value of natural assets? Does it censor the answer? With eyes wide open, let's wander into that lion's den. Since the Fed can not derail us from our career track, we, at least, can make our findings public.

While the US was created to be a democracy by democrats, the Fed was created to be a printing press by and for plutocrats. Our newest central bank took over the power to create new money, a function that the Constitution had charged to Congress. That means the Fed gets to control the money supply. The Fed claims to be a part of the government, which does underwrite it and other global banks. However, during the 2019 shutdown, the government mothballed its websites from lack of funding, while the Fed had its own funds to keep its up and running.

We say "print money" only out of habit. Paper money is for common people in their daily business. Between major players, transactions are usually worth at least 10 figures. For those deals, a big bank taps keys on a computer and – poof! – transfer complete. (An exception was the millions of bills that the Fed of New York printed, wrapped, and sent to the US military in Iraq, which the soldiers used as footballs.)

The Fed says they buy bonds and other paper assets. Yet their buying bears no relation to your or my buying. When you lack the cash yet purchase, first you borrow and go into debt. When the Fed lacks the cash yet purchases, once again you go into debt; you're obliged to pay the taxes that pay off the US bond that the Fed "bought" without money. The money was never transferred from Fed to bond seller but simply appeared in the US Treasury's account. Spending non-money is not counterfeiting, but the Fed did design their notes to look exactly like United States notes, which they are not. What they do is legal but misleading, like the bankers' naming their prize "Nobel."

THE FED INFLATES

The Fed makes it easy to take on debt, public and private, when it extends credit to member banks and the government. The banks create more mortgages. Congress deficit-spends. Since the last recession, the Fed has tripled its holding of US Treasury notes. Government spending has not kept pace, but it has not slowed, either. Without spending a penny – since none of the pennies formerly existed – the Fed over-expands the money supply, resulting in inflation.

Borrowers like inflation. You borrow dollars that are worth more and pay back with plentiful dollars that are worth less. Yet lenders like inflation, too; banks are debtors. The average debt-to-equity ratio for retail and commercial US banks, as of January 2015, is approximately 2:2. For investment banks, the average debt/equity is higher, about 3:1. Repaying with cheap dollars works for banks, too.

When prices inflate, banks must pay depositors more to keep them, and charge borrowers more to profit. When their rates for deposits and loans rise, banks get away with increasing the spread. They also raise their rates on credit cards. For bankers, inflation is a big money-maker.

Knowing this, whenever inflation rises, investors bid up the price of bank stock. When the government announced the CPI went up 1% a month, the same day the stock of the Bank of New York Mellon rose nearly 5%, and by over 10% in one month following the announcement.

Excess-dollars inflates some costs faster, some slower, or not at all. You go to the supermarket and find food costs more, but has your salary risen by a like amount? Not likely. Who benefits? Not the wage-earner. Who benefits are those who can buy low (early) and sell high (later). So, if you're first in line to get the fresh cash, you can make hay while the sun shines. Typically, that's people whose lobbyists stalk the corridors of power and whose money managers trade in big blocks of stock, bonds, and real estate. Of course, the trades must be well-timed, but that's what you have the broker for. Private recipients of the excess money use it to bid up prices – inflation. Since the Fed started replacing US Notes with their own in 1913, the dollar began losing its value at a rate unlike before. When the US stopped "backing" the dollar with gold (1971), then the Fed could print to its heart's content. Americans could not turn to gold coins, being rarely minted, and stamped with a purchasing power far less than the gold's intrinsic value. Inflation since 1971 has soared. Over the past century, the dollar has lost 96% of its former value. In real-speak: The cost of living has risen accordingly.

On occasion, bankers (those typically in charge of the lone legal currency) have issued less new money, which supposedly had the opposite effect and deflated prices, *ceteris paribus*. But things are not all equal. During the Great Depression – an exceptional era – bankers drastically shrank the supply of money and credit. Ever since, the economics profession and its pundits in the press have been able to use the specter of deflation as an excuse for the policy of inflation.

However, a closer look at that Depression yields more powerful factors. Every time, during that lost decade-and-more, the economy showed signs of recovery, the government intervened. FDR's New Deal meant destruction of crops and cattle, tariffs on trade, taxes on investment, curbs on competition between producers, union wages well above commensurate output which entrenched unemployment, etc. Refusing to lend may have been an error but worse ones were made.

Is it wise to fear deflation? Throughout the history of our species, humans are constantly building the better mousetrap. By driving down costs, prices should follow. It's the excessive supply of money that won't let the lower cost of living appear before consumers. In reality, a shrinking supply of money in circulation would suffice in an economy always finding ways to get more from less.

Bankers and politicians say they created the central bank to balance inflation and unemployment by controlling interest rates. However, that method has proven itself ineffective – the economy still exhibits plenty of both; as did inflation, unemployment soared. Further, that method – and the central bank that wields it – is not necessary.

FED & LAND

While he was head of the Fed and one of the most powerful people on the planet ("God on a good day"), Alan Greenspan told Americans that the housing bubble would not burst, that it made homeowners rich. Publicly a cheerleader, privately Greenspan stifled any dissent within the Fed. When the bubble burst and made homeowners poor and mere tenants, and worried investors shifted from stock to bonds, Greenspan's personal wealth (already in millions) rose with the demand for owning the more secure US debt.

When the Fed credits lesser banks in exchange for their bonds, those banks have more to lend to borrowers, who are mainly homebuyers. That means more demand for housing, inflating the price for land, swelling the mortgages. While mortgages still stand out in banks' portfolios, the value of locations can also be part of a business loan or asset management. It's hard to pinpoint how much land enriches banks.

Homeowners like high site-value when selling, dislike it when buying. Realtors and lenders always like it since they gain both ways. Rising land prices not only mean heftier mortgages but also faster turnover, since people are always defaulting or moving in search of a better job.

Hence, the truest value of land is not during the peak – the pre-collapse apex when exuberant speculators engage in a feeding frenzy. Nor is it during recession when many wannabe buyers lack the wherewithal. Rather, it's at some mid-point in the business cycle that paying for land – usually beneath a house – claims around one-fourth of one's income.

To help their ilk laboring away in high-rise office buildings on Wall Street, the Fed "buys" bundles of mortgages from the big banks (much more now than ever), turning the Fed into a de facto landlord. It's also an actual landlord; of the Fed's "assets," over 10% is land – concrete collateral, for a change. How much and how valuable is the land the Fed could foreclose on?

POLITICAL TENTACLES

Sorry, economists are not telling. Thousands of them find jobs within the Federal Reserve. At seven top journals, 84 of the 190 editorial board members were affiliated with the Federal Reserve in one way or another. Journal articles determine which economists get tenure and what ideas are considered respectable. Criticism of the central bank has become a career liability for members of the profession, which silences economists (never a bold bunch to begin with).

While the Fed rides herd on economists, it bestows blessings on insiders. In the year before a presidential election, the Fed tightens monetary policy if a Democrat is in office and loosens it if a Republican is in office. When banks lend less, there's less business, and voters blame the party in power. The editor at the *Review of Economics and Statistics*, a fellow at the Fed, rejected the paper exposing this favoritism, authored by James K. Galbraith, Olivier Giovannoni, and Ann Russo.

A bigger favor than jiggling rates is bailing out big banks teetering on bankruptcy. When Americans were so understandably agitated during the recent recession, Congress felt it had to do something. It went along with Bernie Sanders and Rand Paul and audited the Fed.

Why was a special audit needed? The government has watchdogs. Sort of. As with most of government and big business, there is a revolving door. Regulators spend part of their career in government, part in business, back and forth. Our seeing lobbyists and the lobbied as different entities is only superficial; the elite and the state are actually a partnership.

Until that recession, the Fed had always refused to be audited. To avoid being audited even then, the Fed lobbied Congress. Do other government departments do that? Does the EPA lobby to allow more pollution? Does the Department of Labor lobby to rein in OSHA? May happen.

The audit of Fed books revealed what they wanted kept from the public – how many trillions was their bailout and to whom extended. Even after the audit, the Fed kept doling out credits to their associates – their "quantitative easing". They rescued not just banks but also corporations and not just ones headquartered in the US but foreign ones, too. They didn't stop inflating the value of big business until they'd created and lent nearly $30 trillion dollars. How many dollars is that? If you were to receive them as silver dollars and stack them up then climb them, you'd be on Mars. That's a lot of silver, Tonto.

The recipients of the bailouts found it easy to pay back such huge sums. Getting bailed out confirmed their status as too big to fail, so their stocks and bonds seemed safer to investors. The recipients not only stayed afloat but also swallowed up competitors who were not bailed out (like Lehman), expanding their market share. Recipients used the money they got at close-to-zero interest to buy US bonds which paid a few percentage points – magnified by the huge sums, a tidy profit – all guaranteed.

From the POV of the biggies, while miracles are nice, bailouts are man-datory – a gratifying success for recent recipients. However, for many oth-ers, such as those who lost their homes, the bailouts never tricked down. For the public at large, the recession was a missed opportunity for Shum-peter's creative destruction – sweep out the old bankrupt institutions (lit-erally and figuratively) and usher in the new.

Meanwhile, with all those excess bailout trillions floating around, don't expect inflation to halt or even slow down any time soon. Anyone worried about the deflation bogeyman need not be. Maybe next recession.

SPAWN OF JEKYLL ISLAND

In the interest of full disclosure, let's depart economics for a stroll through political history.

Banks, when functioning well, safeguard savings. When one's sav-ings are many millions – even billions – a local bank is way too small; a global bank fits much better. With investments worldwide and in-fluence internationally, its dividends dwarf its interest payments. Me-ga-banks stabilize fortune, becoming a place to park hundreds of mil-lions of dollars. When big banks control a central bank, risk to fortune is virtually eliminated.

In 1910, the richest financiers of the richest families met in secret on Jekyll Island off the coast of Georgia, not to reform the economy so it would work right for everyone but to curb its mood swings that panicked the very rich. In 1913, the night before recessing for the Christmas break, when few members were still in session, Congress approved their plan to create a central bank. Rather than rename it the "US National Bank," an entity that the popular sentiment had eradicated twice before, financiers christened their corporation the "US Federal Reserve."

Ironically, the Federal Reserve is neither federal nor a reserve. The US president does rubber-stamp each new Chairman of the Reserve and the Fed does pay the Treasury royalties annually. However, the Fed has a cor-porate charter and stockholders (whose identity is debated). Nor does it keep a reserve of its own cash for a rainy day but rather holds IOUs from lesser banks, corporations, and governments, along with smaller amounts of US notes, foreign currency, and gold.

Some of that gold is not theirs. During the Cold War, West Germany thought it safer to store its tons of gold in the basement of the Fed's New York branch. The Fed was not supposed to count it on its asset sheet or use it in any way. In the late 2010's, well after the collapse of the Soviet

Union, Germany managed to repatriate some its gold from the Fed – but only half.

Ten banks control the twelve branches of the Fed:

Goldman Sachs of New York (which supplies so many Secretaries of the Treasury), JP Morgan Chase Bank of New York, Kuhn Loeb Bank of New York, Lehman Brothers of New York (now amalgamated into Barclay's of London), N.M. Rothchild of London, Rothchild Bank of Berlin, Warburg Bank of Hamburg, Warburg Bank of Amsterdam, Lazard Brothers of Paris, and Israel Moses Seif Bank of Italy.

That arrangement between financiers and government did not sit well with everybody. Back in the Great Depression, one congressman, Louis T. McFadden, having served as Chairman of the Banking and Currency Committee for more than 10 years, brought formal charges against the Board of Governors of the Federal Reserve Bank system, the Comptroller of the Currency, and the Secretary of United States Treasury for numerous criminal acts, including but not limited to, conspiracy, fraud, unlawful conversion, and treason. The petition for Articles of Impeachment was thereafter referred to the Judiciary Committee and has yet to be acted on. (Sort of like Congress illegally refusing to call a Constitutional Convention.)

Who are the stockholders in these money-center banks? This information is guarded closely; queries to bank regulatory agencies are denied on "national security" grounds. Yet many of the banks's stockholders reside in Europe.

What's been ascertained is that the American aristocracy still control the Fed. William Rockefeller, Paul Warburg, Jacob Schiff of Kuhn Loeb, and James Stillman of Citigroup (his family married into the Rockefeller clan at the turn of the century) own large shares of the Fed.

The four major US lenders, Bank of America, JP Morgan Chase, Citigroup, and Wells Fargo, own the four major oil extractors – Exxon Mobil, Royal Dutch/Shell, BP, and Chevron Texaco, in tandem with Deutsche Bank, BNP, Barclays and other European old-money behemoths. Those four big US banks are among the top ten stockholders of virtually every Fortune 500 corporation.

OUR MILITARY AT THEIR BECK-AND-CALL

Who benefits from the Fed? With a central bank as a supposed part of the federal government, the eight richest and best-connected families knotted tighter their financial power to the military and diplomatic might of the US government. If their overseas loans went unpaid, the oligarchs could now deploy US Marines to collect the debts.

Marine Major-General Smedley Butler confessed, "I spent 33 years and four months in active military service and during that period I spent most of my time as a high-class muscle man for Big Business, for Wall Street, and the bankers. In short, I was a racketeer, a gangster for capitalism. I helped make Mexico and especially Tampico safe for American oil interests in 1914. I helped make Haiti and Cuba a decent place for the National City Bank boys to collect revenues in. I helped in the raping of half a dozen Central American republics for the benefit of Wall Street. I helped purify Nicaragua for the International Banking House of Brown Brothers in 1902–1912. I brought light to the Dominican Republic for the American sugar interests in 1916. I helped make Honduras right for the American fruit companies in 1903. In China in 1927 I helped see to it that Standard Oil went on its way unmolested. Looking back on it, I might have given Al Capone a few hints. The best he could do was to operate his racket in three districts. I operated on three continents."

Who was Butler? He held the highest rank authorized at that time, and at the time of his death was the most decorated Marine in US history. In 1934 he exposed the Business Plot to overthrow President Franklin D. Roosevelt. He died relatively young at 58.

The bankers who own the Fed lend heavily to munitioneers during war time. They also have a voice in the subsequent peace. At the end of the First World War, Jack Morgan, J. Pierpont's son and successor, presided over the 1919 Paris Peace Conference, which led both German and Allied reconstruction efforts.

The US still backs up big banks and big brokers. A US court, well out of its jurisdiction, ruled that Argentina owed a pair of New York speculators for a debt that Argentina had already paid. In 2012 the US Navy detained an Argentinian warship until Argentina transferred the disputed amount. Bigger scale, since the fall of the Iron Curtain, the US military has invaded Muslim nations. Before the interventions, all of them had several independent banks; afterwards, each had one central bank. Unlike many disparate banks, one central bank gets integrated into the global banking system.

To cement their ties with the military on the personal level, big banks and big brokers hire retired officers.

The facts above by themselves are not damning. The fact that you have to read them assembled here, may be. The story of the Fed shows how insufficient and misleading, perhaps dishonest, it is to discuss economics while leaving out politics.

CRACKING THE FED

Ironically, the US Federal Reserve is not necessary. The market can set lending rates. Competing currencies can reverse inflation. And productive use of land now withheld by speculators can cure unemployment.

It's OK for the Fed to be an umbrella for banks, but to be the lone entity with the power to control currency? And to stonewall inquiries into its bailouts? All could be forgiven if they'd just divulge the worth of Earth in America and the true benefits of ownership privileges.

Petitioning the Fed to release the total value of nature and privilege feels like David vs. Goliath. But if they truly are a public agency as they sometimes claim, they could use their immense resources to calculate and publicize the value of land. But they don't.

The Federal Reserve has outsourced their job (as have many public agencies) of tabulating the value of land. Nowadays they limit their role to being a middleman and introducing curious researchers to the private firm that they, the Fed, uses. For the right price, that firm will provide an answer. Yet if that firm can tally up the rents, the vastly more powerful Fed can, too.

You can see why a programming whiz would be tempted to hack the Fed, though not me. Not because of the rumors that the central bankers play hardball (some theorize they had a hand in assassinations; for me those are dead ends). And not because the Fed has its own oversized and well-equipped police force. And not because the Fed collaborates with the Secret Service, all of which would really chill an interview. But because, first, hacking is snooping, and second, many useful statistics are already available. We ought to be able to cobble together a figure from them.

CHAPTER 27

PORCELAIN ECONOMICS: LOOK, DON'T TOUCH

The First Law of Economics: For every economist, there exists an equal and oppo-site economist. The Second Law of Economics: They're both wrong.

THEY CAN LAUGH, AS THEY SHOULD

In popularity, economist jokes are second only to lawyer jokes (many, blushingly, made up by economists themselves).

> *An economist is someone who lies awake at night trying to figure out how to make reality conform to theory.*

They make it too easy for the jokesters. The discipline fails to account for the worth of Earth anywhere.

WRONGHEADED AND JUST PLAIN WRONG

Paul Samuelson, economist emeritus MIT, had the bestselling textbook of his era, *Foundations of Economic Analysis*, one of the best of all time. In early editions, he explained rent; in later ones, that part was missing. Why delete a discussion on payments for one of the three basic factors in production? It makes no economic sense; oblivious to rent, econo-mists cannot understand surplus or make accurate forecasts. Rather, the decision must've made prudent political sense, appeasing the prevailing winds within the discipline and surrounding it.

To paraphrase proto-geonomist Henry Ford, you can have any statis-tic you want as long as it's not rent. Without the anchor of land, without clarity on principles, without needing to be right, economists can say and believe anything – and they do.

What could be more at odds than the diagnoses and prognoses of liberal and conservative economists? In 1974, two economists – Milton Friedman and Paul Samuelson – shared the bogus "Nobel" prize for say-ing exactly the opposite things.

If all economists were laid end to end they would not reach a conclusion.
—George Bernard Shaw, playwright, social critic, proto-geonomist

In the absence of a key component of the system, spending that never rewards labor or capital, unchecked opinions crowd out facts. Peruse the various explanations by economists re the burst of the recent "housing" (home site) bubble:

- Some blame low-income borrowers.
- Some blame lenders.
- Some blame government pressuring lenders.
- Some blame speculators.
- Some blame regulators.
- Some blame the absence of regulations.

Economists are anything but objective, with no axe to grind; they can be as politically opinionated as anybody. Blaming their favorite fall-guy is where most analysts halt their peeling of the onion. One sees what one expects, and upon finding that does not dig any deeper. On their less than rigorous analysis they base their recommendations. One insider, peeved at the policies of his cohorts, charged them with practicing "flat-earth economics."[1]

THE GAPING HOLE

Can the worth of Earth in America really be known? Economists have not exactly been Mr. Answer Man. Rather, the few who do investigate rent – spending that can never reward effort – have little impact on the rest of their field, who have closed ranks around the leftover topics.

While economists don't count how much society spends on the nature it uses, they don't want anyone else to measure it, either. Like specialists everywhere, the economists, the statisticians, the bureaucrats, act out the Priesthood Syndrome. Within economia, the hierarchy in which they rank each other excludes the uninitiated, the questers after a solid figure for society's surplus.

Peter Preston of *The Guardian* bewails, "When it comes to the economy, nobody knows anything." That might be a stretch; probably somebody knows *something*. Paraphrasing Oscar Wilde, "Economists know the price of everything and the value of nothing."

---

1   "Larry Summers And Flat-Earth Economics" by Steve Denning, *Forbes*, 22 Oct 2015

When it comes to one part of the economy, natural resources and privileges, it seems economists know *neither* price nor value. Even though economies generate a surplus – a fat one, if they'd only measure it (Ch 25) – economists abide by the frame of "scarce resources." Their fake poverty consciousness reinforces society's poverty consciousness. And makes abundance too good to be true.

Without a figure for rent, rent is easily ignored. Overlooking rent, economists ignore the difference between payments for things others create and payments for things nobody created. Doing that closes off from academic view the essential actions of economies, making accurate analyses impossible.

FIELD OF BROOMS: UNDER THE RUG GOES RENT

You read the proclamations of the economists and statisticians and sense no doubt. You listen to them being interviewed or on the phone and hear the voice of confidence. But it is easy to be certain with the backing of the powerful.

Rather than be scientifically correct, economists must be professionally correct. All know that the richest Americans founded the US central bank, and that the Fed and bankers in general are outwardly dismissive toward land. If the most powerful fiscal institution on the planet snubs rent, why should anyone else, especially anyone hoping for a career in the field, research rent? Most don't go there. By de-legitimizing land, politics has discouraged any would-be real scientists.

Instead, academics concern themselves with topics that their colleagues care about – marginal cost, elasticity, etc. And the academic press publishes reams and reams of papers – mere embroidery of little utility. So, less relevant issues become most important and the most relevant issue takes on the tone of "flakey."

While economists show great solicitude toward the elite, some sociologists show consideration toward the lowly. Sociologists used to conduct the wallet test; they'd leave wallets in train stations around the world to measure morality regarding property. The closer to the equator, where people are poorer, the more likely the finder just kept the billfold. The closer to the poles, where nations are developed, the more likely the finder tried to return it (even without accepting a reward). After a few years of testing, sociologists no longer wanted to potentially embarrass poor people. Yet right off the bat, economia chose not to annoy the rich and diligently overlooked surplus wealth.

Economists not only ignore the nature of land – its ability to consume the lion's share of our spending – but also ignore the nature of customs. They treat the sanctity of property, for example, as immutable as the law of supply and demand, a physical law. We deal with laws like gravity by engineering solutions, and deal with customs like, say, gleaning of fields by the poor after harvest, by aiming shotguns and hiring lawyers. If you treat phenomena that can change the same as you treat phenomena that can't, then your conclusions must eventually – or quickly – become out of date.

Students Beg for Relevance

Within the economics discipline, a tyro researcher has a choice. Either adhere to science, or appeal to prejudice. Either perform the most rigorous analysis and follow the findings wherever they lead, or follow the crowd and skirt the true nature of riches.

Most float in the mainstream. Many economic models, and especially those currently *en vogue* in economic theory, suffer from excessive use of mathematical techniques. Young researchers make their models complicated to impress a prospective employer and to get the paper published in a respected journal.

Because economists go through similar academic training, they act like a guild. The guild mentality renders the profession insular and immune to outside criticism. Conventional economists occupy the driver's seat but are headed nowhere near understanding the big picture.

Joe Earle, Cahal Moran, and Zach Ward Perkins in *The Econocracy* report that their lecturers did not mention the biggest economic catastrophe of our times; the recession that started a decade ago. What they were taught had no relevance to people's lives. Students were memorizing and regurgitating abstract economic models for multiple choice questions.

Some professors considered discussing the problem of "inefficiency" in finance to be outside the mainstream. Other academics have some jargon for that (naturally): "cognitive capture." During the financial crisis, cognition staged a jail break. Economists from the Bank for International Settlements (BIS), the International Monetary Fund (IMF), and elsewhere started looking at "inefficiency."[2] We'll see if cognition gets put back in its cage soon; looking at class influence requires *so much* chutzpah.

Insiders Bemoan No Usefulness

All their worrying about controversy makes it hard to look broader and deeper. Focused on permissible topics, economists cripple their field, keeping it from realizing its potential. Instead of of presenting a recipe for prosperity, mainstream economists come up with anything but that.

Within economia, the thinking is screwy enough that even conventional economists feel they must speak up. Over time, economists themselves repeatedly fault their field. Throughout his career, the World Bank's Paul Romer made a habit of critiquing the credibility of macroeconomic models of his peers, irritating many of them (Ch 16).

Romer's academic interest lay in studying how the diffusion of knowledge relates to output; more knowledge, more output. Conversely, less knowledge, less output. The less economists know about rent, the fewer useful insights they can provide society. *"Economize at all costs."* (Get it?)

What do economists have to say about all this? OOH, Prof. David Colander wrote *Why aren't Economists as Important as Garbagemen?* (1991). Irrelevance, *schmirelevance.* Prof. Ariel Rubenstein said academic studies are not intended to be practical.

The gatekeepers of knowledge appear impregnable. Yet, just when it seems that the silencers have won, students revolt. Economic students, academics, and professionals who hope to change business-as-usual within economia have gathered under the banner of Rethinking Economics. It's a hopeful sign, but it's also like a periodic rash. The previous generation had their post-autistic economics. And before them, there were other critics calling for a fresh start.

CHECK YOUR ETHICS AT THE DOOR

"The only reason I don't sell my children is that I think they'll be worth more later."

When Russia struggled to makeover itself and open its markets after the communist state gave up, you could read American economists claiming that the Russian economy was doing swell – even if the people were suffering. That's almost in a league with, *"We had to destroy the village to save it."*

Economists are unusual, but not unique, in their frequency of scoring low in altruism. Adam Grant, in "Does Studying Economics Breed Greed?", notes that even thinking about economics can make us less compassionate. Along with directly learning about self-interest in the classroom, because selfish people are attracted to economics, students end up surrounded by people who believe in and act on the principle of self-interest.

Extensive research shows that when people gather in groups, they develop even more extreme beliefs than where they started. Mob psychology. By spending time with like-minded people, economics students may become convinced that selfishness is widespread and rational – or at least that charitable giving is rare and foolish.

"Ethics teaches us that virtue is its own reward, economics teaches us that reward is its own virtue."

Rather than confront the powerful hierarchy, most economists bow to privilege and sacrifice the scientific method. Still, they boast of being neutral. So, standing by at the evisceration of their field and looking the other way is neutral? Really? Are they neutral or in a state of denial? If not something far less flattering.

Actually, I feel a little sad for the ones who adhere to the catechism. But, at least the pay is good. Still, are we being harsh? What if your accountant did not do the best job possible? Or your dentist? Or your tailor? You'd be in hot water with the IRS, maybe with the wrong tooth yanked, and wearing ill-fitting clothes.

ALL CREEK, NO PADDLE?

Such criticism may encourage economists to change. Wider spread dissatisfaction can help bring about positive change. Even if, so far, there has been no self-correction within Economia.

An early practitioner, Thomas Malthus, called his field "the dismal science." Yet only half of that is true; the science part is a polite exaggeration. Their beliefs are more like superstition. I don't mean to be harsh, but what else can cold, impartial logic lead one to think? Economics is really just a discipline. And its disciples are not full of much useful advice. Rather, they are full of contradictory advice. So, society loses needed guidance.

Peer-review pressures economists to conform. When Gregory Mankiw, a Harvard prof, later a presidential advisor on economics, published a paper predicting the last recession, his colleagues roasted him. Instead of praising his attempt to be scientific (he cited the 18-year period in detailed demographics) other highly credentialed voices upbraided him. Not even the guys at the top can escape criticism when they actually try to do their job.

That chastisement was for forecasting, not for measuring the value of land. For measuring, the penalty is not ridicule but being ignored. Colleagues express no interest – and isolation is not healthy for one's career.

161

Prof. Bill Black, the author of *The Best Way to Rob a Bank is to Own One*, says "economists are unique among scientists [sic] in the frequency, severity, and persistence of their errors. No other field has such a disastrous series of predictive failures in modern times. No other field gives Nobel [sic] awards to economists for preaching critical policy issues and predictions that have proved dead wrong. Yet conventional economists claim that they should be judged on the basis of their predictive success." In an article by Yves Smith at *Naked Capitalism*, 16 May 2016.

Predict is what a science can do. It's the acid test of whether or not a study is scientific. For a field to belong to the noble endeavor of science, its practitioners must be able to predict. Economists cannot predict with any certainty. Ergo ... economics it's not a science.

Economists have called eleven out of the last nine recessions.

The *Farmer's Almanac* does a better job forecasting the weather. Astrology puts economics to shame, joked John Kenneth Galbraith, the world's best six-foot-seven economist and President Kennedy's Economic Advisor. Two other guys who together, back in the 1990s, won that ersatz "Nobel" nearly brought the global financial house down by persuading the rich and powerful to invest in their hedge fund that went belly up.

Q: What do you get when you cross the Godfather with an economist?

A: An offer you can't understand.

The rare economist does find some predictive power; *"residential investment leads the business cycle whereas non-residential investment lags."* That's by Morris Davis (Georgetown University) and Jonathan Heathcote (Federal Reserve Board) in "Housing and the Business Cycle" (Ch 13). They even mention land and acknowledge it's absent from NIPA or GDP, so they whittled it down to 20% of the price of housing. For academia-nomics, at least that's something.

Are we finding too much fault? We're not. The key to forecasting accurately is not a secret, but it is ignored. Leaving out rent, most professionals miss that this cycle occurs with depressing (pardon the pun) regularity, making prediction possible. Geonomists see that, much to their advantage. Rents are vastly superior tea leaves, as you'll see.

CHAPTER 28

A SOUND ALTERNATIVE LEFT IN LIMBO

"Allow yourself ... to be amazed by the mystery of the mundane."

– Paul Heyne, *The Economic Way of Thinking*

L and is out of sight, sut of mind. Even most above-average people still don't get it. The brain's hardwiring leaves nearly everyone ill-attuned to the worth of Earth in America. Nearly everybody assumes land value is like any other value that we seek when buying or surrender when selling. We don't realize that it comes not from anything the seller has done but from what society has done – densify population – and what nature has done, like provide a great view. We're in denial that natural rent is money for nothing.

Statisticians don't tell us how much we spend on the nature we use. What economists do tell us about the biggest stream in the economy is unfocused, confounding forecasters and optimists. We can't tell how good we have it. Those who get rent can ... then pull up the ladder, focusing attention elsewhere. Critics and reformers busy themselves fighting their own small-picture battles.

A blind spot of another species that leaves me dumbfounded is the one afflicting those lizards who live on the cliffs in Chile overlooking the Pacific Ocean. They fight among themselves for the best locations to build a nest, but don't fight against seagulls when the birds come to eat the lizards' eggs. The parents sit there and watch placidly, clueless as to what the devouring of their eggs means.

We humans don't let predators eat our kids but do watch without understanding as members of our own species slurp up surplus, doubling the mortality of newborns.

The practical-minded turned land into real estate, the idealists into environment. Watching land disappear makes a gadfly feel invisible and a little sad. Being so relentlessly inquisitive cuts the quester off from the rest of humanity. Such is the life of scouts on the cutting edge, those who explore the wilderness that nearly no one else is interested in yet. It's a lonely pursuit without much in the way of real-time rewards.

EVERYBODY GUARDED

Bringing up such payments that never rewards anyone's contribution of labor or capital can make recipients tune out the insight.

The biggest obstacle to sensing rent is not the 1% covering up the role that society and nature play in generating such a huge surplus of value. It's The 100%. Everyone's interest aligns to obscure land and its profit.

Most economists declare land to be "irrelevant today" and a theory of rent "simplistic." Yet Ford Foundation professor Dani Rodrik noted we must simplify the world to understand it. Conceive the simplest possible explanation, argued Albert Einstein, an admirer of Henry George. George, the 19th c. economist, land reformer, and author of the classic best-seller *Progress and Poverty*, based our getting of goods on axioms of human nature.

Rather than curious about rent, economists are incurious about the difference between spending that rewards another's effort – paying for goods and services – and spending that never does – payments for land and privilege. And, typically, economists treat privilege as a given; as natural law, as immutable as gravity.

Other fields have professional standards, why not economists? Even lawyers have professional codes of conduct (for whatever they're worth). How about a pledge to insist upon the most precise economic measurements of the most elementary spending flows?

GEONOMISTS – LET THERE BE LIGHT

The discipline of economics can be fixed. Just insert rent. Then economists could take a stab at making accurate predictions and turn their discipline into a true science. Using the research of Homer Hoyt, at least three geonomists predicted the recent recession:

- Dr. Fred Foldvary in the US at San Jose State University, cited in *Harvard Faculty Insight* (by Teo Nicolais, 18 Oct 2016), wrote in 1997: "the next major bust ... will be around 2008 ..."

- Fred Harrison, PhD and reporter in the UK, and

- Phil J. Anderson in Australia, author of *The Secret Life of Real Estate*.

Well in advance of the recession (Foldvary a decade in advance), they explained why it had to happen and when it would happen. Ironically, these geonomists who can predict don't win much media attention while economists who can not predict do win the scholarly prizes.

Those geonomists were successful because they watched the clock – the 18-year land-price cycle. As society spends more for land nobody made, it must spend less for goods and services humans did make, until a recession results about once every generational period. Demographers cite the rise and fall in population growth over the same period.

If hidebound economics discipline can't be fixed, it can be replaced. Like going from astrology to astronomy and from alchemy to chemistry, the next shift of the paradigm could be from economics to geonomics. F.A. Hayek compared the economy to an organism, Troy Camplin to an ecosystem. Actually, observed Herman Daly, formerly of the World Bank, economies are a subset of the ecosystem. All three are getting close to recognizing the economy as the geonomy.

Geonomics is the field of study that finds patterns and natural laws in our producing, distributing, consuming, and protecting of goods and services, all interacting as an organic system that "inhales and exhales." Since economists won't, geonomists do admit who does the work and who gets the wealth. Since economics won't, geonomics must answer how economies function, why regularly they fail, and what must change to fix that.

While the status quo might relax – and even celebrate – re nearly everyone's out-of- touch-ness, I'm feeling a tad sad for the economists who're presently denied the chance to become experts and scientists. Their physics envy shows they long to become real scientists. If geonomics does supplant economics, many academics might be happy to have the long nightmare of crippling the study be over.

A Generation Ready?

Thomas Kuhn in his *Structure of Scientific Revolution* showed that fundamental new ways of doing things come not from within a discipline but from without. What lies outside mainstream economics that could upset the current apple cart? Geonomists? Perhaps a new generation? Kuhn explained old minds don't change, only young ones are open to bigger picture explanations. An economist is someone who didn't have enough personality to become an accountant.

Geonomists would observe the wrong triggers of youngsters and avoid them, and figure out their right buttons and push the latter. Open links of communication. Then titillate curiosity, open eyes, and enthuse new agents of change with a new way of seeing the world. Clearly state how the enthused can pitch in. Certainly, some of them would lend their help to push Leviathan to measure our spending for the land and resources we use.

And maybe just in time, if money for nothing is a major driver behind our widening inequality, despoliation of the environment, politically-charged non-solutions, and widespread resignation and fatalism. Whether we're on a tipping point or not, glimpsing a way out of our economic morass is a glimpse worth having. Eh?

CHAPTER 29

AS CRISES CLIMAX, WHY TALLY RENTS?

First enjoy, last do what works. What are priorities?

CULLING ALL SOLUTIONS

Many scientists say certain problems will reach their tipping points within our lifetimes. Humanity meeting its needs has caused: loss of farmland, loss of drinking water, fouled air, cancer epidemic, obesity, unaffordable healthcare, unaffordable housing, exhaustion of the employed, low self-worth for the under-employed, disappearing jobs, income lagging behind inflation, indebted governments and college grads, endless wars for resources – not to mention awful daytime television (still – and it's prime time). You can easily add to the list.

Seeing these as crises has kicked millions of people into gear. They want to save the world and save it now. However, they're going about it like a wild goose chase, with little recognition of any underlying cause. Focused on their surface solution to their chosen problem, they've not peeled the onion. Their vision has not penetrated to the fundamental flaw.

Yet when many things go wrong, they typically have one root cause, noted Henry George long ago. Thanks to locations acting like geysers, a few lucky owners, lenders, and investors reap extra profit without adding extra value. As profiteers go about capturing those rents, they create problems downstream. Publicizing a figure for rent highlights this casual connection. Hence reformers would welcome the statistic (Ch 4), and, before their issue worsens.

ECO-LIBRIUM UPENDED

There is one issue that is totally indifferent to anyone's political beliefs – our economies' assault on Earth's ecosystems. Lately climate change gets the publicity and may merit it, but global baking is actually one assault among many. Just one: the background noise teaching children to tune out, making adults live in constant denial the norm. Then every breath taken, every drink gulped, every morsel eaten is to some de-gree toxic – unless we buy bottled water and organic produce. Yet it's hard to avoid the sluggish traffic jam, escape the smudged horizon, and breathe cleanly, deeply.

Machine exhausts – especially from cars – cause <u>allergies</u>, <u>asthma</u>, and autism. Farm runoffs and discarded medicines poison the water table. The toxins intentionally lathered upon cultivated food, acid rain, plastics and other garbage choke the oceans, devastating fisheries and the production of oxygen. Some power plants and weapons tests add to background radi-ation which mutates living cells.

Modern economies pollute merely as a byproduct but *pollute* profuse-ly; already two of five will get cancer and the rate keeps rising. Modern economies *deplete* in step with population growth; of course develop-ment gobbles farmland, construction destroys forests, factory farming erodes topsoil and expands deserts; those three developments together endanger species and drive essential links in the food chain into extinc-tion. Future hunger looms over the shoulder of modern obesity.

Environmental pollution – from filthy air to contaminated water – is killing more people every year than all war and disease in the world. More than smoking, hunger, or natural disasters. More than AIDS, tuberculo-sis and malaria combined. The number of people killed by pollution is undoubtedly higher and will be better quantified once more research is done (same story for the worth of Earth).

So long as humans profit, whether using or abusing land, the health of the planet must worsen. Singly or together, these environmental issues are closing in on their tipping points, putting civilization on the brink of e-collapse. Afterwards, you would not want to be anywhere near the sur-face of the Earth.

Housing Out of Reach

Some people, even rich ones, spend 40% or 50% or more of their in-come for a home (the kind sitting on land, unlike dirigibles). Some low-income workers double up yet still sacrifice too much on shelter. Meanwhile, zero-wage beggars camp out.

While they call it a "housing crisis," that's superficial observation. It's actually a location, location, location crisis. Housing ages and wears out; it depreciates. It's the land underneath that appreciates. What newcomers to enticing areas will pay dearly for is less the building, more the climate (California) and/or culture (Portland).

You can blame techno-progress, giving raises to its workers, endowing them with the wherewithal to bid up homes on desirable sites. Billionaire venture capitalist Peter Thiel explained, *"the vast majority of the capital I give to the companies in Silicon Valley is just going to landlords. It's going to commercial real estate and even more to urban slumlords of one sort or another."*

> *I didn't become a software engineer to be trying to make ends meet.*
> – a Twitter employee

Furthermore, when mega-cities like San Francisco, New York, and London force entrepreneurs to focus on making money just to pay the rent, that stifles entrepreneurship.

Those who flee spendy regions create a ripple effect, making other places unaffordable for residents there. People who can't afford to live in San Francisco and Boston, etc, settle in Portland and Buffalo. There, location values rise, displacing people to Boise and Harrisburg. When people move out, they take their "effective demand" with them. They relocate their land values from their old locales to their new ones, padding values there. Hot spots like the coastal cities radiate rent that warms up formerly cool spots.

If jurisdictions had a solid figure for the value of their locations, residents would know how much revenue might be available for housing help.

RECURRING RECESSION

People called the last major downturn a crisis, a Great Recession, forgot about it, and yet will call the next one a catastrophe, a true depression. The Kondratieff Wave could be a factor, but definitely the bailout. Last recession, the Treasury and the Federal Reserve created $30 trillion by issuing it. Big banks and big business, domestic and foreign, used it to buy stocks, bonds, and real estate. They bid up their price beyond their value, exposing asset owners to bankruptcy besides putting assets out of reach of most people.

As long as speculators bid up the price of land beyond what most can afford, the US economy must recede on a regular basis. As people spend more for never-produced land, they spend less for the goods and services their neighbors produce. As producers sell less, they hire less. After a while they must fire people. In the words of three academics writing for the Federal Reserve, *"in the Great Recession ... the collapse in the housing market was followed by a sharp rise of unemployment."* And the unemployed

169

don't make the best of consumers. Nearly the entire economy recedes. The recession lasts until loan defaults knock land costs back down again (for a while).

You could see it coming – as did the geonomists – if you tracked the fluctuations in the flow of money that society spends for the nature it uses (Ch 28).

With land *et al* being out of reach, with nothing to fall back on, the next recession will crush more people.

Many people accept hard times as natural events, like droughts, which human ingenuity cannot deter. Of course, various thinkers have proposed various solutions, mainly focused on banks. However, for a solution to succeed it must address both kinds of spending, the one that rewards production versus the one that rewards rent-seeking. Attend to that *sine qua non*, then the business cycle will no longer boom and bust but climb and glide.

Uncle Sam Bankrupt

Per capita, US public debt is currently (updated regularly) $62,000. The US debarks the fourth most tourists annually (70 million), unloading dollars abroad. Further, nations with high inflation hold many US dollars in reserve; places like Ecuador even use it for their national currency. With those two escape valves, the Federal Reserve's over-issuing of new dollars inflates domestic prices by "only" 3% (officially). That still doubles prices in 25-30 years. However, it's too slow for most people to notice. The frog in the saucepan on the stove.

Meanwhile, the private debt of citizens is also enormous. Most of that debt is mortgage, and a good half of mortgage is land value. Credit cards were in second place. Now it's student debt – $1.5 trillion; while current law allows governments to declare bankruptcy, it prevents students from doing the same. Facing few well-paying jobs, future generations could stay in serious debt their entire lifetimes. Lifelong debt must make it look like older generations pulled up the ladder.

The third big bookkeeping category, corporations, is also a swamp of debt – $29t (just coincidentally, the same amount as the Fed's '08-09 bailouts). Often, they borrow to grow, pay it back, and write off the interest payments. (Did bankers have a hand in lobbying for that loophole, too?) Without that money-saving write-off, businesses would borrow less, and the overall economy would not be saddled with so much debt.

But tax deductions don't do much good when there's not much income to deduct from, and there's not much during recessions.

During recessions, people demand help from government, but can't pay as many taxes. So, governments sink even deeper into debt. At some point, just as Detroit did, the US must go bankrupt. To pay its debt it will devalue the dollar relative to other currencies, just as the UK did with the pound. Then others will abandon the dollar, flooding the US with its own money, driving inflation out of sight.

If the US is to skirt that scenario, it must peel the onion down to its core. Become bold enough to confront speculation in land, which inflates the cost of borrowing. If we knew the value of land, that would put a spotlight on ground trafficking and its unwanted offspring.

1% – OUT OF CONTROL

While economies can't help but create a surplus society-wide, enclosing that abundance creates a dearth in some pockets. The few who collect rent use it in ways that constrict supply and exacerbate demand. They buy up (thus bid up) assets and under-utilize prime sites. With higher prices and lower wages, many people cannot afford to own large assets, such as a home. Homeownership has fallen and stock ownership was never a majority pastime. Both behaviors drain away the wealth of the majority, making them more vulnerable. It widens the wealth and income gaps, which slows real economic growth, and increases poverty. U.S. income inequality, on rise for decades, is now highest since 1928.

The abyss of inequality – it's unimaginable. A few get so much that they could not spend it all in several lifetimes of ultra-luxurious living. And their locations rise in value faster than elsewhere occupied by non-rich. Meanwhile, others barely scrape by on the leftover scraps. Beyond the purely material dimension, concentrating so much wealth and income in the hands of so few is bad for everyone in many ways. Even people who aren't poor feel like losers. Wildly unequal societies are less happy than more equal ones.

The more disconnected from others one feels, the less one cares about others, and feels comfortable committing all sorts of anti-social behaviors. The current occupant of the White House may be an outlier, but most of the 1% share his egocentric aloofness. The insensitivity of the privileged crosses party lines. While conservatives might oppose safety for employees at work, liberals might not hear the poor oppose public housing because they, the poor, don't want their offspring trapped in it, generation after generation.

Expect the next recession to create more losers and winners, further shrinking the anxious middle class, expanding the lower class, while making the upper class more aloof. Then America stops being a predominantly middle-class society. That means losing its democracy, losing its environmental movement, losing its rapprochement between the classes. Perhaps it's the nature of democracies to devolve back into aristocracies in a feudal society. How many recessions away is America from morphing into something even more like George Orwell's *1984*?

It's ironic, happening now – as other big nations are growing their middle class: China, India, Russia, Brazil. Just give those trends enough time – ours shrinking, theirs expanding. ("The American Middle-Class is No longer the World's Richest" by David Leonhardt and Kevin Quealy in the *New York Times*, 22 April 2014).

Then, no way America remains the flagship nation among democratic global powers.

CRISIS = OPPORTUNITY? AT ALL?

Nowadays, even middle-class residents spend over half their income on housing (that sits on desirable sites). In most US cities, downtowns are pockmarked by vacant lots that, if developed, could hire hundreds of workers. And in Louisiana and other cancer alleys, pollution greatly suppresses land value.

How much worse can it get? How much more can the tattered social contract tolerate? How much more can the ravaged global ecosystem endure? While such trends seem to move at a glacial pace from the perspective of the individual, when they reach their tipping point, the collapse comes quick – as did the USSR. By then it will be way too late to take the relatively easier steps of prevention versus the arduous steps of curing catastrophe.

Wannabe reformers have much bigger fish to fry than tallying rent. And boy, are those fish ever getting fried! Like generals believing they know how to win the last war, the left tries to rein in business, the right tries to rein in government. *"The definition of insanity is doing the same thing over and over again and expecting a different result."* – supposedly Einstein, an admirer of proto-genomics – but more likely from a 12-Step Program.

Environmental organizations, trying to stave off eco-lapse, are not necessarily stuck in that old Industrial Era conflict of boss versus worker. Yet, most act like economists in keeping "neutral." Property is so controversial that they don't touch it. Further, many environmentalists are already homeowners, which is to say beneficiaries of site values going bonkers. Overlooking the linkage between Earth's worth and Earth's health is in their self-interest, too. They, too, turn a blind eye to the link of reward-to-ruin – the role of rent in driving eco-ploitation – and confine themselves to "just say no" to despoliation.

People interested in new ideas and old problems try to solve these problems without addressing the basic motive – possessing rents. You can see how that's worked out. So, who's really rearranging deck chairs on the *Titanic*? We gadflies or wannabe saviors who studiously ignore the worth of Earth in America? If our business-as-usual is to be altered in planet-saving ways, and *in time*, a critical mass must become cognizant of the real role of rent.

CHAPTER 30

DO WHAT OTHERS WON'T? NAIL DOWN RENT?

Alfred E. Neuman: What, me worry?
So, if they won't worry, should we?

NOW WHAT? QUIT, OR KEEP AT IT?

Hear them? Noisy posterity becomes an irritating taskmaster, insisting present peoples correct their economies. A lot is at stake for future generations, as each current generation fouls its nest and battles with each other.

Most of what afflicts us is not only misguided economics. Nor is it just politicians passing into law shortsighted policies. It's also our narrow-minded customs regarding "something for nothing," etc. (Ch 2).

Solving all that depends on addressing one obscure issue – determining how much we spend on the land and resources we use. That lone datum, if widely known, works wonders.

- Investors and savers can find better forecasts (Ch 28).

- More people can see how bountiful economies are (Ch 24) and lose their poverty consciousness.

- Prosperity helps de-motivate our senseless mistreatment of one another. The future could breathe a sigh of relief.

CAN'T COUNT ON THE COUNTERS

Those who *can* calculate the total won't; no expert or official source has supplied a realistic estimate for the worth of Earth in America. No mainstream economist uses the statistic in their theories nor as an economic indicator. Despite reliable prediction being the essence of science, they ignore geonomists who do forecast accurately (Ch 20).

Nobody should ever have to apologize for their curiosity, yet some in the discipline question our quest. They bring up objectivity and impartiality, like people in glass houses hurling stones. Their stifling behavior is maddening; it's what Melville's *Billy Budd* felt, being mute, unable to be heard. Eventually it drove him insane. Fortunately, the road to that sad outcome is blocked to us by you, our dear reader.

If someone did the hard work, perhaps a solid stat for all of society's spending for nature and privilege could motivate a critical mass of researchers to upgrade, to geonomics. Such things happen in the realm of science; eventually, Leviathan can be turned. The old beliefs that generate psychological dissonance no longer comfort the mind. Overcome resistance to the new paradigm and the once-ridiculous becomes mainstream. Most new ideas begin as heresy.

CAN'T COUNT ON THE REFORMERS

There are a thousand hacking at the branches of evil to one who is striking at the root.
– H.D. Thoreau

While our problems are dire, our activists flail about blindly. If ever they are to solve chronic and pressing problems, agents of change have to see the big picture. Part of that picture is rent, the surplus that an economy naturally and unavoidably cranks out. Who that stream rewards, and what they do to capture it, shapes the world.

Environmentalists, for example, want our species to befriend Mother Earth. Yet it's not rational of them to let land be such a fat profit-maker and expect a law that *"just-says-no"* to development to succeed. Ordinary citizens are moved by the bottom line, so the planet needs profit on its side. Then, even if for no other reason but to save money, most would spare ecosystems.

A figure for how much we spend on the nature we use helps other problem solvers better see how economies operate and when they don't, what to do differently. Then we could adopt policies that redirect rent so it'd no longer reward waste but only efficiency. Again, the public would save and befriend reform.

CAN COUNT ON A GADFLY

The public has a right to know the amount and could put the knowledge to good use. President Theodore Roosevelt not only endorsed the public recovery of land values (a little historical footnote), the "Roughrider" also inspired his readers: *"It is hard to fail, but it is worse never to have tried to succeed."* And when the going gets tough, the tenacious gadflies get going.

Recall Bertrand Russell's conjugation: *"I am stalwart. You are stubborn. He, she, or it is pigheaded."* You can see where they slide me in, and where I place myself. Philosopher Lord Russell, the person who scored the highest ever on math in the entrance exam to Cambridge without ever previously going to school, eclipsed Teddy by proposing that government pay the citizenry dividends from the recovered rents of sites and resources (another historical factoid).

Once you know that rent rules, how can anyone let go of this quest? The few who grasp that our spending for land and resources creates problems now, perhaps solutions later, are the ones who must press forward and drill all the way down to the most serviceable number for the size of all rents. We'll see this through to the end. Go over the heads of the data-keepers. Issue a rallying cry. Intertwine counters and demanders. Determination feels so much better than despair.

An Endgame Strategy

If on my watch things worsen drastically for humanity, it won't be because I failed my responsibility to make rent knowable. Once we make an intriguing figure available, we'll have laid the groundwork for specialists to refine it. We'd have an authoritative statistic for all spending for all kinds of land and privilege.

Some open-minded agents of change are curious to know how much society spends on the land and resources it uses. And some cutting-edge researchers are willing and able to calculate rent's role and size. Even if no deep-pocket person or foundation steps up, in this e-era of sites like Kickstarter, money may no longer be an obstacle.

No worries. We'll *un*earth the worth of Earth in America. Plus Privilege, too.

CHAPTER 31

NEW HOPE FROM NEW STATISTICIANS, INC.(S)

"Being a statistician means never having to say you're certain."

OUTSIDE-THE-BOXERS DEMAND & SUPPLY TRUE STATS

A few on the cutting edge of economics may soon address basic statistics, on both the demand side and the supply side. New outfits have sprung up capable of ferreting out a figure for the worth of Earth in America. Plus, some new organizations may demand a grand total of the value of land, resources, and privileges. So, what if the Bigs have not yet come through. These Littles could fill the breach.

Not that we should give up on the Bigs, even though the official sources of statistics have not made basic research any easier, and the entrepreneurial sources have not made their statistics any cheaper. But at both places their personnel is constantly changing, so their degree of cooperation is constantly in flux. Perhaps some competition from the Littles may nudge them in the right direction.

The few big-name economists who criticize official statistics and demand better numbers are not alone (Ch 17). To meet the need for accuracy and relevance, already, lesser known, specialized organizations churn out stats and make the data available to inquiring minds. Given their self-described willingness to think outside the box, might such entities calculate the economic surplus which the presence of society generates?

POTENTIAL SUPPLIERS – FIRMS FOR REFORMED FIGURES

Here are the Top Ten groups that supply relevant statistics – stats that you cannot easily find elsewhere – who may find intriguing the challenge to measure how much we spend on the nature we use:

1. Geodemographics Knowledge Base (GKB) is a comprehensive directory of hand-selected websites for people interested in the application of geodemographics and geo-spatial analysis. Sounds so close to geonomics, it could work.

2. Statista's team of trained specialists analyze, audit, and update each of their statistics frequently. They attend to detail and adhere to academic archetypes and guarantee keeping to high standards. They claim to be fully equipped to meet the needs and expectations of their users. Of us, too?

3. Zero Hedge takes its name from the quip by Keynes who said in the long term, survival drops to zero, so don't hedge your bets on your lifespan.

Their mission:

• widen the scope of financial, economic and political information available to the professional investing public;

• liberate oppressed knowledge;

• provide analysis uninhibited by political constraint; and

• facilitate information's unending quest for freedom.

Sounds like a winner, eh? The next, a UK variant, looks promising, too...

4. New Economics Foundation (NEF) is the UK's leading think-tank, whose mission is to kick-start the move to a new economy through big ideas and fresh thinking.

They do this through:

• high quality, ground-breaking research that shows what is wrong with the current economy and how it can be made better,

• demonstrating the power of fresh ideas by putting them into action, and

• working with other organizations worldwide to build a movement for economic change.

5. The Centre for Research on Globalization (CRG) is an independent research and media organization. The Centre acts as a think tank on crucial economic and geopolitical issues. Its articles, commentary, background research and analysis focus on social, economic, strategic, and environmental issues.

6. The *Washington Post* has its Wonks blog which focuses on the economy among other subjects. Writers there break down issues and make their arguments with statistics. They explore issues in depth; one of them even writes about land topics!

7. Shadow Government Statistics ("Shadow Stats") by John Williams offers analysis behind and beyond government economic

reporting. Despite Williams coming from the mainstream, there are an awful lot of mainstream articles criticizing him. That actually could signify that he's telling the harsh truth that the powers-that-be don't want the public to hear. May he go broad and deep and tackle the worth of Earth.

8. Washington DC's first progressive multi-issue think-tank, the Institute for Policy Studies (IPS), has served as a policy and research resource for visionary social justice movements for over four decades. Barbara Ehrenreich, author of *Nickel and Dimed* (2001), who describes herself as *"a myth buster by trade,"* once worked there.

9. G. William Domhoff, who goes by "Bill," is a Research Professor at the University of California, Santa Cruz. Four of his books are among the top 50 best-sellers in sociology for the years 1950 to 1995. *Who Rules America?* (1967, Ch 12) was loaded with statistics and described the local growth machines of real estate development.

All these outfits (except the next) are open-minded enough to go beyond the official numbers. Nine of them raise the expectations of curious geonomists, number 3 especially. Of them all, the best known is the least bold – #10 (coming up). You know all the oddities you find in statistics (usually negatives, like plane crashes or suicides bunching together)? There's money to be made citing those.

FREAKS TO FACT-CHECK THIS?

10. Freakonomics was launched by Stephen J. Dubner and Steven D. Levitt. They're academics, like Dr. Domhoff, but despite the great name, the "freakonomists" stick to the mainstream, unlike Bill Domhoff. Their "-omics" began as an article, then a book, then morphed into a documentary film, a Jon Stewart show appearance, and an NPR radio show. Freakonomics.com has been called "the most readable economics blog in the universe" (which, the Steves admit, isn't really saying much).

In 2014, Levitt and Dubner published their third book, *Think Like a Freak* – supposedly a blueprint for an entirely new way to solve problems. However, they use official statistics with blind faith. Like the rest of academia, they call their numbers "data" (though it's nothing like the accurate and relevant measurements in hard science).

Dubner and Levitt focus on funny things people do – like *When to Rob a Bank*. These little-picture oddities – not how economies work and don't – they call that economics. Their legion of followers, who number in the millions, also think fun-with-numbers is real economics. Give the Steves their due for making economics entertaining and enlightening, even if less relevant.

All the probing into the unusual by the Freaks has, so far, kept them clear of the rent factor. Limiting "economics" to the incidental, while eschewing big-picture analysis of an economy's actual workings and its resultant wealth, does the discipline no favors. Their analyses elevate what's merely entertainment, distracting the public, and trivializing what the public should know, cementing their institutionalized indifference to deeper analysis.

Maybe another source is the fact-checkers. Nowadays, some Internet divers (not surfers), organizations, and newspapers pride themselves on testing official numbers for veracity. With the rise of the Internet and googling, fact-checking has become big business. Perhaps one of them would check Larson's, Albouy's, or our total for America's land value.

FactCheck.org is a project of the Annenberg Public Policy Center of the University of Pennsylvania. The APPC was established to create a community of scholars within the University of Pennsylvania. It addresses public policy issues at the local, state, and federal *breadths* (not levels, a misnomer).

Once an official estimate of the social surplus that land-value represents comes out, then let doubters fact-check that.

POTENTIAL DEMANDERS – THE RICH

May this soloist effort of ours play matchmaker and introduce those whose mission it is to supply insightful data to those who could put out a figure for Earth's worth to good use. There are a few groups already interested in land, in rents, and in statistics.

First, believe or not, some rich folks want to share the wealth – not *all* the wealth, but some. They may not go so far as to advocate changing the system that made them wealthy. But they do want to close the income gap at least somewhat.

- Bill Gates's dad wants to tax very rich people.

- Warren Buffett's son wants to require charities to spend all their money on charity, so that they can no longer perpetuate fortunes.

- Likeminded top one-percenters have banded together to form groups like Patriotic Millionaires and Smart Capitalists for American Prosperity.

In the past, some of the wealthy have even bit the hand that fed them.

- The US has a Communist Party due to one of the inheritors of the General Motor's fortune funding its supporters.
- An inheritor of a tobacco fortune campaigned against the sale of cigarettes. And ...
- An inheritor of the Baskin-Robbins ice cream fortune tried to persuade Americans to adopt a diet that left out sugar.

Perhaps it's self-serving – they wish to stabilize the system that put them so high above everybody else, since yawning disparity de-stabilizes the social order. Or maybe it's merely an embarrassment of riches. Or those elite are being charitable out of the goodness of their hearts; philanthropy is a normal human trait. Whatever the motive, it's exciting to know that people with the power to make change are advocating change. Some even advocate fixing or replacing GDP to measure an economy's success; a figure for rent could make a great replacement statistic, updatable daily.

Rather than share just any kind of wealth, maybe these elite would warm up to sharing rent, once they hear how society generates it and how much it is – enough, if shared, to make a real difference. Rent is the kind of wealth that's less a product of individual effort and more a result of social progress and natural advantages. By tapping sites, resources, and privileges, we can avoid taxing the successes of useful producers, and thereby not diminish the incentive to produce.

MORE MAYBE DEMANDERS: RELIGIOUS, COINERS FOR THE REALM, YOUTH

Second, some of the faithful talk up land issues that cause suffering: concentration of land ownership, homelessness, unaffordable housing, pollution, etc. Many recent Roman Catholic Popes have urged followers to reform land holdings and to quit contaminating the earth. The Catholic Worker Movement in the US and Oxfam in the UK might lend their voices to a chorus calling for officials to measure the Bible's "fruit of the earth," which belongs to all.

Third, people bothered by the fact that dollars are no longer backed by gold propose returning to gold, or backing with some other valuable tangible, like land. Whether that's a practical idea or not, its proponents might first need to know how valuable land is. Certainly, they would join the call.

Fourth, youth is a time when humans tend to be more open to new ways to see the world. Inexperienced. Ahistorical. Impatient. Idealistic. And lacking an intellectual lineage they're deeply emotionally attached to. Note that it's grad students who criticize economics. And youthful activists who latch on to cutting-edge causes. With the youth wing of monetarists, religions, and wealthy families leading the way, demanding answers, then suppliers of measurements would come up with a good, useful number.

For academics, always in need of young blood, it must feel odd to regard the coming generation as a potentially disruptive force. Some professors would not want to be out of step with their students. Those young-at-heart profs would sniff around the issue of rent and echo the call for a serviceable number for the value of land and resources.

DEMAND SPURS SUPPLY, RIGHT?

Once a critical mass of economists requests decent measurements of spending on assets never produced, then the bureaucratic part of government would raise their bar. Once enough reformers demand to know the size of social surplus, then the legal system itself would push up the bar. The data-minders would make their statistics as accurate and relevant as possible.

You may want to ask any of these organizations for a reliable figure for the rental value of all land and resources in the US or UK. Your question could add to earlier ones, putting those interrogated at a tipping point, where they will take on the challenge of digging deeper. Or you may find by then they'll have come up with an answer and you'll be one of the first to learn it.

The past successes of the above interest groups give one grounds for believing that a gadfly can win. Such groups have expanded their horizons before and can do so again. Once cognizant that our spending for things never produced is a social surplus, probably they'd seek to know not just its size but also the identity of its recipients.

CHAPTER 32

MODERN ENCLOSERS HINDER COUNTING THE RENT

My shotgun terrified every picnicker in the park; guess the place is mine now.

FIEFDOMS FOR FREEHOLD ONLY

Countrysides are where the fewest people own the most land for the least money. Some billionaires in the wide-open West own spreads the size of small East Coast states. Their holdings make for an incredible overall concentration. The USDA figures 3% (not a typo) of Americans own 95% of the privately held land ("Land Rush" by Peter Meyer in *Harper's,* January 1979). That's a far cry from the ideal of proto-geonomist Thomas Jefferson, who thought America should be a nation of small farmers.

Other ranchers would like to own more and set their sights on public lands. Some have christened themselves the Sagebrush Rebellion. Opposing them are the democrats (lower case – those who favor group participation in decisions), who resist losing public property, even if they're unaware of the public nature of rent.

Angry propertarians focus on land, not on rent, since rent *is* the source of their fortunes; who wants others poking that deep into their business? By making their claims strident, they make tallying the worth of Earth in America awkward. Not just because the claimants are rich and powerful; they also have the myth of rugged individualism – now a customary defense mechanism – on their side.

LAND SANS DUTY

While militating against paying the public for their land, sagebrush rebels and other rural corporations are not shy about taking public dollars. Agri-business gets most of the federal billions, but also lining the trough are ranchers, loggers, and miners. States and localities also chip in in their own way, with roads and tax breaks.

Land-grabbing gentry cast themselves as working ranchers defending their way of life. In actuality, they tend to be absentee owners with multiple homes, one a fashionable address in town. And if they are purely country, I bet their lawyers are citified.

Cities are where opposition to landlords is most strident, given unaffordable housing and gentrification. Many of those who hang on to their apartments demand rent control. Aspen, Colorado-ites won a tiny land tax to fund public housing. Less conventional people are resettling in land trusts, more every year.

While proclaiming their exclusive right (actually, privilege; rights are inclusive) to own any and all, land grabbers reject any requirement to be good stewards. Their straw-villain is the environmental movement. Now that many citizens, mostly city dwellers, want to protect the environment are these "allodials" (owners owing nothing) bitter.

The propertarians picked a winnable fight. As audacious are the allodials about taking over the earth, that's how timid environmentalists are about sharing the earth or her worth. Since most of the latter are home(site)owners in metro America, they too benefit from the current system inflating site values.

Despite that tactical advantage, the belligerence of propertarians makes sense, since claims to land are tenuous. Legal experts acknowledge that titles to land are never thoroughly clear. As they say, go back far enough and you'll find that all titles are based on force or fraud.

SHARED SPACES – ZILCH

The more extreme rentiers agitate to abolish all public property and any vestige of the commons. Their ideal is reached when private individuals owns the roads, national monuments, beaches, et al. Sidewalks would be gated – if any were to remain and not be ceded to automobiles. Ultimately, would the 1% own the rivers? The lakes? The atmosphere? All are for sale.

No shared spaces at all? If we are left in shock and awe at their audacity, know that this stance of the wealthy anti-communitarians runs gratingly against the grain of normalcy. However individualistic people may think they are, most citizens enjoy parks and the wilderness.

The loss of commons would likely accelerate the loss of community, further atomize humans into dust motes in a cloud, detached from all others. Lacking trustworthy relationships, when troubles arises, it'd be Hobbes's war of all-against-all.

That may be farfetched, but the loss of civility and tolerance is not. Community is the context for morality. We learn our ethical lessons from the family, friends, and residents around us. The less community, the less common morality.

Some of those who have benefited the most from what civilization has to offer oppose not just commons but society itself. During the 1980s a certain politician across the pond claimed, *"there is no society"* (may she RIP). On this side of the Atlantic, a TV actor/politician (may he RIP) argued government is a problem, not a solution.

No Shared Power, Either, Thank You

Government agencies themselves can share the same philosophy. Rather than meet their mandate, they outsource their services to corporations, such as private penitentiaries and banks. Supposedly it's to cut costs and control the economy.

Those who oppose government take aim mainly at public goods and social programs, not the police or military. However, if government lacked the force of arms, it'd hardly matter what laws they passed or what rulings judges gave or what fines the IRS levied. They'd have no way to enforce them.

Meanwhile, contracting out the tasks of government has not downsized government. The state has not been withering away but expanding, whether beneficial or not. This expanding government sometimes oversteps its bounds:

- the IRS hounding people who owe nothing into bankruptcy,
- a neighborhood losing its very existence to a city-backed developer,
- a judge with "black robe disease" finding innocents in contempt of court.

Not exactly user-unfriendly.

Not surprisingly, victims of such abuse find appealing the ideas of shrinking and privatizing the role of bureaucrats and politicians in their lives. OTOH, citizens who've never experienced mistreatment and/or need their government jobs, equate government with social cooperation. They see anti-governmentism as anti-social madness.

Those opposed to government had better beware of what they wish for. Soon the only state left could be what it once was—the gentry's expanding incarceration, law enforcement, and the military. An anti-authoritarian's opposition to those policies could become hazardous.

Rural rentiers typically lack respect for "eggheads." Perhaps the rift can be employed to win for gadflies some recognition from academics, as long as their rules are followed. Using their conventional definitions and methods, not corrected by reason or deep analysis, then what would a total for the worth of Earth in America look like?

CHAPTER 33

IF LOWBALLING INPUTS, HOW MUCH IS RENT?

If you can't beat them, arrange to have them beaten.
– George Carlin

FOR THE SAKE OF ARGUMENT, SURRENDER

In recalling how privileged landowners can voice anti-social views yet wield so much influence (Ch 32), I'm still reeling. The wealthy have made it a jungle out there for academics and bureaucrats. Economists tread lightly around the interests of those powers behind the throne (Ch 12).

Academics and statisticians tend to lowball the value of land and resources (Ch 18). They come across not as partial to the truth, but partial to a special interest; for them, a high estimate is high risk. So, let me reprogram myself to not just appear impartial but go one better and appear anti-partial. Forget logic. We'll lowball, too, and see what that total of all rents looks like.

Sigh. Surrendering to convention feels like throwing a robe on a classical nude sculpture – dressing up a figure for the worth of Earth in America in pseudo-respectability. Yet it may be easier for a science popularizer to propagate a total that's pseudo-official, no matter how far off market value it'd be. We twist the campaigner's saying to yield, *"I don't care how much the total becomes as long as you spell socially-generated value right."*

TRADE LOGIC FOR CONVENTION?

We gadflies tried seeking an official total for how much all of us spend on the nature we use. But none's to be found. Official statisticians don't do their addition in the rent room (Ch 25).

Was it Gandhi, who said everyone has a piece of the truth? Economists, however, have *pieces* of the truth. Rather than tabulate a grand total for our spending that never rewards anyone's labor or capital (nobody-made land), they've measured parts of the whole, like residential land (Ch 13) or parkland (Ch 15).

Not finding an official total, we sought, instead, an accurate total. That meant we questers had to question the statistics and methods that experts used. Normally, questioning implies criticism, and who enjoys being criticized? The experts have not exactly been effusive with their praise, toasts, and calls to celebrate this quest to know the total of all rents (Ch 19).

If you can't beat them, bury them – beneath opaque jargon and a thicket of insignificant statistics; that seems to be the motto of many specialists (Ch 17). Our contrary modus operandi was to dig up the numbers that most fully represent surplus – the worth of never produced sites and resources. Yet any gain in accuracy has a cost – a loss in credibility within mainstream minds. Our conciliatory motto is the more familiar, "If you can't beat them, join them." That is, base our extended quest on where officials left off.

Now, what we lose in accuracy may mean a gain in acceptability. A total born of their own assumptions may placate the specialists. This exercise could win friends and influence people toiling away in academia or in a relevant bureaucracy. They may become more forthcoming with their assistance and feedback. They may even take on a pride of ownership.

Or not. If the figure is still too robust, it may leave too much flesh exposed for a prude loyal to the pack. Ironic, eh? A too-high figure scares away conventional specialists. A figure too low and they say rent's too small to bother with. So, what's their Goldilocks figure?

Pragmatists put public acceptance above letting one's light languish beneath a bushel basket. Idealists, of course, put accuracy above accommodating those who don't prize truth. Our curiosity resolves the dilemma. Exactly how far apart are the two totals – the one catering to convention versus the one employing informed logic? Let's see.

INPUTS UPDATED

If economists focus on land at all, it's land out of sight, underneath homes (Ch 13). As they should. Spending on housing drives much of the economy, and the land component accounts for much of the economy's booms and busts.

- The latest mean for housing is $203k per house in America. That's by Case & Shiller, academics cited by government and media. Or, it's $218k, according to Zillow, a company that serves homebuyers and businesses involved mainly in building houses. Multiply that $15k difference by 83,000,000 homes (Census Bureau stat), you get $1.2t. While specialists might be comfortable with a difference of a trillion here, a trillion there, anyone seeking a reliable total for land value must keep looking.

• Figures for housing hint toward a figure for land. For residential land we have academic Lincoln who, for the start of 2016, put it at both $8.7 trillion (FHFA) and $9.9tr (Case & Shiller again). Both totals cannot be right. Indeed, both can be wrong. The difference is $1.2t (again). That much money could provide a comfy income for one-fifth of all US households. Such disparity makes one wonder if all conventional figures are wrong. Almost all give more weight to buildings, which depreciate, than to locations, which during most of the business cycle appreciate. Nevertheless, both figures can serve as a minimum for all land price.

• The only academic figure for land based only on land sales (not sales of location and structure together) is Albouy's, for metro land. His 2009 total was $18t. To extrapolate a figure for 2016 (Lincoln's latest), use the FHFA stats at Lincoln; the percentage increase in land value from 2009 to 2016 is 33%. So Albouy's 2016 total comes to $24t, only for urban land.

• To add the price of rural land, the USDA says the price of all farmland alone in 2016 was $2.8t. To add some value for all ranches, mines, oil wells, water, parks, etc., you could probably double that amount. You'd go from $24t, through $26.8, to $29.6.

• The only official figure for all land of all uses is Larson's $23t for 2009. He uses those official estimates, which are low. To increase his total for the start of 2016, multiply by the one-third. That bumps Larson's total up to $29.6t – what we found using Albouy.

• To add the value of nature's electro-magnetic spectrum (that we use for modern communication), it was at least a half-trillion in 2007[1] and certainly more a decade later, bringing the total to over $30t. That amount was reached in 2006 before the bubble burst, and supposedly land values had already recovered a few years back, so $30+tr likely underestimates the actual total.

To keep extrapolation to a minimum, we'll leave out the trillions due to (a) utility monopolies, (b) environmental license (official tolerance of

---

1   "America's $480 Billion Spectrum Giveaway" by J.H. Snider of New America Foundation, July 2007

degradation) and (c) privileges like corporate charters which limit the liability of polluters and depleters.

Since this aggregate $30t is price, which is derived from rent, and rent conveys the realistic picture, we must convert price back to rent. In Ch 20 we used 10% of price. Lowballers prefer 5%. That puts the quasi-official total rental value for all land in America at a mere $1.5tr

Is that $1.5t credible? It's less than one tenth of national spending (or income or GDP). Consider the value of locations in popular cities. Consider the commercial value of downtown sites. Consider the spending on oil from domestic fields, on owning frequencies in the EM spectrum. Consider imputed environmental values. From such a bigger-picture POV, no way can the rental value of nature in use be $1.5t. Indeed, the totals that geonomists came up with, from $4 to $6t, are far less partial, much more logically derived, and hence far more likely to be accurate (Ch 20). Yet for the present, precision is not our goal; acceptance is. Can mainstreamers live with this paltry figure?

Conformist economists may still ignore rent. However, a total of whatever size that should gain traction with the discipline is one from a source above reproach – business. Then statisticians and politicians will have to deal with that.

CHAPTER 34

FIRMS OFFER A RENT TOTAL – FOR A PRICE

Pay the piper, call the tune; but will it be melodious?

ANSWERS FOR SALE

Astronomer Carl Sagan's book, later a film, *Contact*, got NASA to finally fund SETI. (signals from outer space are also ignored, just like sightings of land value, so the chapters of this book are in good, celestial company). If it becomes as successful as Sagan's, this book (maybe later a documentary), could get agencies and foundations to fund the determining of a new indicator – the worth of Earth in America.

Some well-connected companies - self-described experts at tabulating – do sell measurements. They claim the ability to calculate whatever feature of an economy or demography that a buyer could want. So, that includes figuring out the value of land and resources? And provides a total more precise than our current ballpark figure?

Firms like Case-Shiller, CoStar, and Statista suit the needs of the mainstream. Case-Shiller are cited all through the business press, CoStar sells to the Federal Reserve, and Statista draws from over 20,000 sources. These three offer to sell the total value of US residential real estate, and property values in general – more than we want.

With their resources, they may also be able to determine just the well-cloaked land portion of real estate. And they may be able to tack on the value of land used for purposes other than residential or commercial or industrial: i.e., agricultural, pastural, sylvan, mineral, and spectral (the airwaves), and other resources, such as water. Asking them to figure in the value of anything else natural may be asking too much.

Whatever they'll do for you, they'll do it for a price – affordable or not.

MONEY MAKES THE GROUND GET COUNTED

Good data costs big bucks. That's one good way to keep the curious hoi polloi ignorant – raise the price of knowledge. For a researcher or a writer conducting a labor of love, that cost is an out-of-pocket expense from not-so-deep pockets.

Who besides a gadfly researcher will pony up? Presently, the specialists who suspect the relevance of rent won't invest their talents (Ch 25). Those who could afford to hire investigators don't sense the relevance.

Even the generous who're interested in economics, like most everyone, have a hard time seeing the relevance of rent. One spending stream can't be that significant, can it? What's the big deal about rewarding producers of goods and services, versus rewarding owners of never-produced land and resources? Not much. Unless you want to forecast, assess your surplus, and project the gains from healing land.

Raising Funds

Given the millions of human minds and the diversity of human interests, some individuals and foundations do get it; those are the geonomic (Earth-focused economics) ones. The activist group Common Ground USA, comes to mind, as do two New York-based foundations: Henry George School and the Robert Schalkenbach Foundation. Even this writer has on occasion been funded by them to research rents.

In the NGO world, the above are small fry. But there are well-heeled sources to target:

- foundations with deep pockets set up to benefit the public by giving grants to enable the necessary research that no one else is doing; organizations with large memberships who respond to special appeals, and

- well-heeled individuals known to donate to worthy causes.

In the era of modern social media, millions on the cutting edge are sympathetic to crowdfunding appeals in Kickstarter, Indiegogo, GoFund-Me, *et al.* At some of these sites, it takes money to make money – they charge fees. At all of them, it takes time to learn the ropes.

Fortunately, some humans are curious enough about the true worth of Earth to give money to a stranger to find it out. A critical mass wants to know how much they and all their fellow citizens spend to own or use a location, or natural resource or government-granted privilege. Plus, they could claim credit for launching a new and better indicator; and for making it safe for funders and researchers to climb aboard the rent train.

Money in hand, one could hire the pro statistician. Yet, needing to turn to a for-profit corporation raises the question: why should taxpayers be funding the responsible public agency? Why spend taxes on those huge staffs in public agencies just so they can pay huge staffs in private agencies? If a public agency won't perform its fiscal duty, why not abolish it and save the public money? An agency fulfilling its mandate – is that something to lobby for?

CHAPTER 35

MOVEMENTS FIND A PIECE TO THEIR PUZZLE

If only they knew now what they'll know then …

ACTIVISTS FRY FISH UPSTREAM?

Try as hard as academics and bureaucrats might, it's nigh impossible to do economics without getting politics all over you. If you find that competition accomplishes some goal equitably, the gallery calls you a rightist. If you find that cooperation accomplishes some goal efficiently, your detractors call you a leftist. Comes with the territory.

While economists try to steer clear of politics, political people grab their favorite economic policy and charge up the hill with it. Which could work out for us gadflies. Some of those political people have influence and may help us sway officialdom to calculate the worth of Earth in America. They'd do that because such a statistic would help them advance their own agenda. Remember those issues at their tipping point (Ch 29)? They're still tipping.

With the ability to inform their listeners about how much society as a whole spends on land and resources…

- Environmentalists could extrapolate how much more valuable healed nature would be;

- Urbanists could dangle a fat plum before reformers of local revenue policy;

- Fans of full employment could show what constrains vs creates opportunity;

- Income re-partioners could grasp what widens the gap and may close it;

- Libertarians could show the feasibility of freedom from conformist work;

- Businesses could present an alternative to taxes on business; and

- Governments could demonstrate how successful they've been.

Advocates in all these movements might join the call to officials to calculate the total of all rents.

GREEN OVER GREY

Buying land or anything with land in it, we reward owners, some of whom chew up the environment and spew out pollutants. Rather than continue to enrich those who waste resources or leave behind waste, society could make them pay. To avoid paying an extra expense, producers and investors would switch from foul, inefficient ways of providing a good or service to clean, efficient ways. Businesses and residents would reduce their depletion and pollution, too.

Just in time, governments may make polluters and depleters pay. Government could collect Ecology Security Deposits, require Restoration Insurance, auction off Emission Permits, penalize violators of standards with realistic fines, etc. To charge any of these, government must first know the market value of affected regions and surrounding regions. That's our cue.

Another policy that benefits the environment is one some jurisdictions use in order to raise revenue efficiently. They shift the property tax from buildings to land. To pay the land levy, owners of prime vacant lots put them to good use. Then downtown absorbs new building, leaving little or no construction to sprawl over the countryside (Ch 39). Before more localities levy land, they'd like to know its yield.

Environmentalists defend hillsides, creek banks, and marshes and argue for zoning and other regulations. However, the only current way to profit from land is to develop it. An alternative way could both reward the owner and spare the land. That is, government could gather up the annual rental value in its jurisdiction and pay residents a share. Receiving their portion, owners prevented from developing their land would be compensated financially, and get to live in a vigorous environment.

A region with parks, open space, and wildlife corridors has higher value than one with wall-to-wall development.[1] So does a region that humans have cleaned up. The cleaner the environment, the more valuable the locations,[2] and then the bigger the share. Residents have less motive to exploit their land in any way that'd lower its value and their share of rents. Talk about a positive reinforcing feedback loop.

1   "Why America Needs More City Parks and Open Space" by Paul M. Sherer for The Trust for Public Land, © 2003
2   "Cleaner Air Results in Higher Home Prices" by Matthew Davis, NBER, March 1999

New Urbanists

Many longer-term residents hate to see their town grow too popular and densify. "Welcome to Ourtown; now go home," NIMBYs say. Yet as long as human population grows, so must demand for human habitats. More residents push up site values, add to traffic, etc., etc.

In response, urban advocates aim for two major goals: affordable housing and walkability. Both are hard to win when locations are astoundingly pricey.

Developers cannot profit by putting low-cost housing on exorbitant sites, parcels that carry high property taxes. Residents respond with requirements to build affordable housing, tax breaks for developers, tax caps, etc. Ironically, the effort to pay less to government means paying more to sellers. Speculators simply raise prices, absorbing any tax savings. Look at California after Prop. 13 and the states that followed suit – site costs are out of sight.[3]

While capping the property tax does not work, shifting the property tax does. Now taxing property penalizes improvers and rewards speculators, but it could be flipped to discourage speculators and reward improvers. Doing that addresses both issues, affordability and livability (Ch 39). To enjoy these benefits, cutting-edge urban advocates support shifting the property tax.

Furthermore, to free themselves of confining social norms, people who long to express their uniqueness move to cities. Amid the population density, they find both anonymity and others of their kind. They find the freedom to be themselves.

Jobs Chase Workers

With automation performing more jobs each day, many humans are worried. Most of them qualify for income only by performing a job. When jobs disappear, so does their income. So, they propose to conjure jobs, even if they're not productive but merely conformist, like bureaucratic desk work. What a way to waste a human life.

Ironically, at the same time the desperate beg for jobs, their politicians tax them. Taxing wages makes hiring workers more expensive, so firms don't. Conversely, de-taxing jobs increases job opportunity – albeit by itself not enough for everyone wanting one.

Along with taxes, another job-blockage is speculation in land. Some investors and owners make money by not using their land but by waiting

---

3   "California Real Estate Median Prices of Existing Homes since 1968" at RealEstate ABC, 15 December 2015

to sell (or lease) it at a higher value. Even in Manhattan, where a parcel can make one a fabulous fortune, you see vacant lots, abandoned buildings, and under-utilized locations. What happens on a vacant lot? Nothing, at least in the way of work. Take a bird's eye view. Where the city is pock-marked by vacant lots – slums – that's where unemployment is endemic.

Those eyesores, when located downtown, displace businesses and residents to less central, less desirable locations. There companies make less money, hire fewer workers, and pay lower wages. And wherever they are, vacant lots attract crime but do not foot their share of the bill for police and fire. These unwanted results of vacant lots, along with the argument that society generates land value, are more reasons to tap rent to benefit society.

The IMF (International Monetary Fund) notes that, having to pay some sort of land dues, owners put their vacant lots to good use,[4] which requires hiring people. Afterwards, using the new structures requires hiring people, too. Further, where employment rises, so do wages. Furthermore, where wages rise, so does location value. Were it measured, policy-makers would know its potential to do good, and job-pushers would be one big step closer to their ideal of full employment.

LIBERTY LOVERS

Modern libertarians claim yesteryear's classical liberals as their intellectual lineage. And even though many libertarians see paying land value to community (instead of to a bank) as a tax and taxation as theft, and receiving a share as a handout akin to a bribe, they are out of step with their roots. Libertarian heroes Adam Smith, Thomas Jefferson, and Tom Paine, and others all stated that spending for land – our mutual heritage – is something that all members of society would be better off sharing.

Along with lionizing past liberals, modern libertarians prefer their governments slender (as do freedom seekers of all eras). The revenue policy of sharing rents can grant their wish. Jurisdictions that recover the rents of land, such as Hong Kong and Singapore, need not tax labor or capital. The jurisdictions that pay rent dividends, such as Alaska paying residents the oil dividend, need not subsidize citizens, whether rich or poor. Presto. Governments regain their youthful figures.

Ironically, it was Alaskan Libertarians who won that oil dividend and the libertarian organization the Heritage Foundation that ranks Hong Kong and Singapore as two of the freest places on Earth, while remaining mute about their policy of capturing rents. Most modern libertarians are

4   "Taxing Immovable Property: Revenue Potential and Implementation Challenges" by John Norregaard, IMF Working Paper 13/129, 2013

too anti-government to be aware of the libertarian nature of socialized rents, but not all. Some rich libertarians and libertarian groups are on board, too. Others could join the call.

While another world war is not pending, all the smaller wars that the US has confined itself to are troublesome. So are the black budget and the heavily-armed planet. The more a nation knows inequality, the more belligerently it behaves. Conversely, the smaller the prosperity gap between nations and between classes, the more peaceful are the nations. And rent, once counted, could become a bigger pie for society, better divided.

CURRENCY CRAFTERS

Not so much a movement as a discussion group scattered nationwide, are wannabe reformers of economies; a passionate segment focuses on money. Among those, some would abandon the call to return to gold or any precious metal as a backing for newly issued currency. Instead, they'd use the value of land. If that makes sense, then they and their audience might want to know land's value.

BUSINESS

Another non-movement, but immense component of society wielding enormous influence, is business, from local Chambers of Commerce to multi-national corporations.

Most of them profit more from capturing rent[5] than from returns on their inventiveness and delivery of desirable goods and services. But no segment of society is monolithic.

There are some businesses who realize that putting prime land to good use attracts investment. After Pittsburgh widened the gap between its half of the property tax on land and its half on buildings, the city renewed its downtown without one penny of subsidy; rather with private investment. Savvy business-people who take note of results also understand that it's on prime sites where one can make the most money.

Such people have on occasion endorsed shifting the property tax:

- Nationwide, the Multi-Family Homeowners Association, who are the owners of apartment buildings,

- In Portland OR, major developer John Russell, and

- Believe it or not, in one year both the Greater Philadelphia Association of Realtors and the Chamber of Commerce.

---

5   "The hidden rent-seeking capacity of corporations" by Prof Dirk Loehr in *International Journal of Social Economics*, Vol. 41 Issue: 9, 2014

Businesses tend to be less willing to leap before looking. You'd think they'd like to know the size of their new land dues. They could use their influence to win an estimate for the value of land.

JURISDICTIONS

A few politicians have endorsed public recovery of socially-generated land value. Some state reps:

- In Michigan, a candidate and some office-holders;
- In California, Mike McGuire; and
- In Pennsylvania, Wayne Fontana.

And decades ago, presidential candidates of the two big parties – Democrat Edmund Muskie and Republican Jack Kemp.

In years gone by, back when the idea of using land value as public revenue was *the* progressive policy, even sitting US presidents voiced approval, including Democrats FDR and Grover Cleveland and Republican Teddy Roosevelt. Louis Brandeis, a Supreme Court justice, was on board.

Across the pond, two British prime ministers, Winston Churchill and Lloyd George, pushed the shift of the property tax off of buildings, onto land.

More than just talk, several members of the Oregon legislature have introduced bills. Elected reps did the same in Maryland, Minnesota, Missouri, New York, and Virginia. Connecticut passed legislation to allow certain cities to test out the tax shift; to date, none have.

Longer ago, several jurisdictions actually adopted the policy. Every place that has implemented it has benefited, in terms of higher wages, slower inflation, greater investment, gentler business cycle, wider spread prosperity, etc. For others to follow suit and take it mainstream, they might like to first know the size of their tax base.

LOSING DISSONANCE, MAKING COMMON CAUSE

Nowadays, even middle-class residents spend over half their income on the housing (that sits on desirable sites). In most US cities, downtowns are pockmarked by vacant lots that, if developed, could hire hundreds of workers. And in Louisiana and other cancer alleys, pollution greatly suppresses land value.

In the past these movements—from environmentalists to politicos— have been at cross purposes, which leaves society in a somewhat turbulent state and even the victor feeling a bit uncomfortable. To lose that social dissonance, some advocates in disparate movements welcome the opportunity to work across the aisle with opponents. Perhaps tabulating the natural surplus of our economy could become a mutual goal.

The above interest groups have expanded their horizons before and succeeded often. Singly or in a coalition, they could go over the heads of data-keepers, straight to their bosses. The lobbied officeholders would order the public's number-crunchers to make an accurate accounting of society's spending on nature and privilege; imagine: an agency fulfilling its mandate.

CHAPTER 36

PEOPLE POWER TO DE-CLASSIFY DEEP DATA?

Ask officials for a stat for Earth's worth. If that's too hard, ask the universe.

Not just gadflies are stonewalled when trying to squeeze a response from government; even presidents get stymied. John F. Kennedy expressed amazement upon discovering how little power the President of the United States actually had, how unresponsive bureaucracies were to his executive orders. Perhaps since then, the "highest" office in the land has accrued more power.

RULERS BOSS BUREAUCRATS

Of course, bureaucrats did not stonewall every command issued by their commander-in-chief. Most often, they obeyed in one fashion or another, even to an order pertinent to our request. Throughout history, bureaucracies have complied with their ruler's mandate to calculate the worth of Earth in their jurisdiction: in ancient Sumer and Egypt (Foreword) and medieval England (Ch 3).

How hard can it be? In Denmark, they used to include the value of the location with each number and address in the phone book. Denmark had a history of assessing and taxing land dating back to the Age of Enlightenment (late 1700s). Then, a nephew overthrew his uncle, the king, just so *he* could sit on the throne and tax land for the betterment of the whole kingdom (the idealism of youth). No vestiges remain.

The first US Government, in operation during the rebellion against Great Britain, operated according to the first attempt at a constitution, the Articles of Confederation. That document directed the federal government to fund itself with land value collected by a tax. To do so, of course, required government to know, and thus assess, the worth of locations.

The Founding Fathers junked their first constitution with its land tax, in favor of the current constitution with free trade between states and tariffs at national borders, plus land sales to fund the federal government. While no longer needing to know the value of locations, that document does nevertheless require Congress to conduct a census every ten years.

Today's census counts many things; not just the number of people but also their incomes and outgoes and possessions. The US Census Bureau even used to add up the value of land in America, but quit doing it. Yet, the US Congress could require that department to get back to tabulating the worth of Earth in America.

The Census Bureau is not the only federal agency that Congress could order to calculate the rental value of land and resources. In the past, many agencies have published a report on the total rental value of some, or all, of everything:

> Bureau of Economic Analysis, Bureau of Land Management, Census Bureau, CIA (believe it), Congressional Budget Office, Congressional Research Service, Departments of Agriculture, of Commerce, Federal Communications Commission (for the value of the EM spectrum), Federal Housing Administration, Federal Housing Finance Agency, Freddie Mac, Fannie Mae, General Services Administration, Government Accountability Office, Housing and Urban Development, Labor (Bureau of Labor Statistics), Library of Congress, Office of Management and Budget, Treasury (Bureau of the Fiscal Service), and two quasi-governmental agencies, the Federal Reserve and the National Bureau of Economic Research.

That's a big bowl of alphabet soup. If asked to take a stab at it again, let's hope this time they standardize their goals, definitions, and methods, so that they arrive at a unanimous answer.

For Us: Freedom of Information Act

In a democratic republic, people have a right to know how much their society spends on the nature they use. Under the Freedom of Information Act, any citizen could request the figure for the total of rents. But that assumes at least one official agency has tabulated it.

It also requires that the petitioner understands the obtuse text on the agency's websites on how to use the FOIA (they offer a second website to explain the first one). A citizen can only hope that their request is made according to the bureaucracy's rules. And that, if any mistake is made, it won't completely derail the request.

In total, 20+2 bureaucracies got our own FOIA request; I held my breath. Within days, the letters from the bureaucracies did begin to pour in. Many were form letters, yet many others were unique and original, in order to address my unheard-of request. I was grateful for the time that a busy bureaucrat took to reply.

The end results? I should be honored. I was treated like JFK. Not all even bothered to answer. If an agency did not have the answer, it'd be easy for them to say so. But if an agency did have the answer, it could be hard to say so. The powerful insiders who now capture most of the rental stream have subtly mandated government to overlook that fatness, the very existence of King Rent, in order to keep the public in the dark re this source of fortune. Would that explain why some agencies did not reply?

Those that did reply said they did not track "rents," our society's spending on land, resources, spectrum, etc. Those that did track real estate values said they did not keep separate the value of human-made buildings from the value of nature-made land. Nor did they offer to separate for me the two drastically different values, to make clear our two drastically different spendings – one that rewards human effort (production) and one that rewards social advantage (ownership).

Go Over Their Heads

If our public bureaucracies don't have the answer, elected officials can make them determine it. The infrastructure is in place. Members of the House, the Senate, and the Executive all have their own research service. And when it's a member of Congress asking for an answer, that bureaucrat feels enough heat to generate some light, however dim.

Officeholders might issue such an order if they, in turn, were ordered. Upstream from politicians, a citizen could request their elected representative to require the relevant public agencies to either release a reliable figure for the worth of Earth in America or to get busy and calculate it. That is, if the chain of command in a democracy were to behave as advertised.

Of course, the elite could accomplish that; they can make the law and they can break it (which is the way all societies operate). For an ordinary citizen, it's much harder to persuading elected officials or bureaucrats. Still, citizens asking their reps to ask their public servants is worth a shot.

In communication with an elected representative, various talking points can grab their attention. A letter could look like this:

re: Official Figure for Land & Resources

Hon. [congressperson];

Given your clout with federal bureaucracies, you could be of immense help to the electorate, ferreting out data. Our sought-after

203

figures tell us what phase the business cycle is in (decades ago Hoyt discovered the 18-year land-price cycle). Good to know if you want to keep your savings or investments safe. Those same statistics also tell us how much surplus our economy is generating.

The surplus is our spending for land and resources and other assets not created by anyone's labor or capital. During the Civil War, the US included land values in the tax base. If government could assess location value then, it could do so now, too.

When government does recover "rents" (technically), i.e., puts them in the tax base, landowners use land more efficiently, which generates jobs. Even if the federal government forgoes these rents, it could negotiate with states to levy a tax or charge a land-use fee or institute land dues to direct this common-wealth into the public treasury.

Many voters prefer a more equitable disbursal of our social surplus, rather than let just a tiny fraction of us hog the most. A fair share would finance free time for every citizen. Presently Singapore cuts taxes on efforts, taps into the socially generated value of locations, piles up a revenue surplus, and pays citizens a dividend.

At your beck-and-call you have the Congressional Research Service, the Congressional Budget Office, and the Library of Congress, among others. In the past, economists in such public agencies have taken a stab at coming up with the correct answer. They can do so again, hearing from you.

If no other agency wants to do it, the IRS could. Instead of always prying into one's private wealth, the agency could tabulate society's public values. Performing that service would make the IRS less repugnant to ordinary citizens.

How much we spend for nature can be derived from many sources, including mortgages, property taxes paid, certain insurances, assessments, appraisals, for all uses, not just residential, but also commercial, industrial, agricultural, sylvan, mineral, fossil fuels, the EM spectrum, and not just private land but also nonprofit and public land, including roadways and water. We're talking trillions here, so we really need to know this stat.

There is a precedent. In the past, President Thomas Jefferson (a physiocrat) ordered explorers to bring back from the Louisiana Purchase all sorts of knowledge. Your researchers could bring back a total for society's spending on nature.

Feel free to use any of the language in this letter. If you have any questions, please ask. If a researcher has any questions, I'm happy to answer them, too. If they find themselves unable, we can tabulate

the total, once supported by a fair contract.
I look forward to hearing from you.

Thanks sincerely, John Q. Public

Asking officer holders difficult questions runs into the same obstacles as asking bureaucrats. The politicians, or their staff person who handles incoming mail, may have a radically different worldview. They don't see what the big deal is about knowing the size of all rents. Plus, like almost everybody else, they're busy.

Even though elected public officials are often about as helpful as the un-elected public officials, politicians are here to serve us. Sometimes they do. They are cautious, they are weathervanes. One thing all politicians do is count noses. They won't build a bandwagon; but will hop on one with enough voters onboard. The greater number of people who want to know, the happier politicians are to help. They're human and humans like to be on the winning side. When an election is coming up, incumbent candidates are especially helpful.

Instead of sending a letter that might fetch a form reply, gadflies could make an outing of it. Assemble a crew and pay a visit to their Congress-person; even recruit an academic expert to join in petitioning Congress. In person, they could all insist that public agencies measure the socially generated value of locations and make that knowledge public. Present a sample letter the elected official could use.

LEADING THE WAY

Rational officeholders [sic] should bite. Heck, location value could be a tax base, for gosh sakes. Whether to write a letter or set up an appointment, there's the link to find your federal reps.

Even then, after fielding a request from a Congressperson, bureaucrats might stall, as they did with JFK. Words-on-paper are one thing; adherence to those words is something else altogether. Whatever factors influence an agency's decision to comply, or not, are hidden from public view. Even if the venture is not immediately successful, at least it will plant seeds that will bear fruit in ways one cannot expect.

While waiting for the powers-that-be and specialists to perform, we gadflies can do our own digging and show the specialists the way. Show them all what an accurate answer would look like. And how to find it.

Never doubt that a small group of thoughtful, committed citizens can change the world; indeed, it's the only thing that ever has.

– Margaret Mead

205

CHAPTER 37

THE GREATEST STAT ON EARTH: ITS REAL WORTH

I'd rather be vaguely right than precisely wrong.
— J. M. Keynes, via Carveth Read

LAX VS. LOGICAL

As in the country-western song, researchers have been looking for a tally in all the wrong places. Statisticians have sought a figure for the worth of Earth in America by looking at sale prices, not at rental leases. Compounding their problem, most economists (except Albouy, Ch 13) looked at property, not at land alone. Why?

It's practical. Sales greatly outnumber leases, deals for buildings greatly outnumber deals for raw land, and housing in the GDP greatly outnumbers all other kinds of real estate and businesses based on land like farming or mining. Easy peasy.

Yet you don't do science by choosing easy over hard. You choose what yields accurate results. That's what we'll do.

WHY PRICE BIAS?

While aggregate price is not functional – imprecise and infeasible– total rent is. Yet economists choose the non-functional and eschew the functional. Not for the first time and not likely to be the last.

The media, too, ignore rent, focusing on price. When the price of housing goes up, the media tell us that is good news, while forgetting to point out that actually, it's the location that appreciates as buildings depreciate. Nor are higher prices good for everyone. What's good for the goose (the seller) is not necessarily good for the gander (the buyer).

And why do we see the seller in the driver's seat? The price for land is not how much the seller asks but the final amount that the buyer negotiates. It's the demand of and competition among buyers that sets price, far more than the stubbornness of sellers.

Economists and the media reflect popular longing. Everyone wants to sell out at an exaggerated price and denies the reality of needing to next become a buyer, likely at an exaggerated price. The prevailing mindset of the typical American is speculator, not money-saver.

Pulled by statistical ease and pushed by cultural norm, almost all of those few economists who do guesstimate the worth of Earth in America used price – a dead end. And when presenting a value for an asset, the default figure for economists is from the POV of the seller, speculator, profiteer, the "winner," the recipient from the deal; not the buyer, investor, payor, the "loser," the expender in the deal. In politics, winners write history. In economics, profiteers command statistics.

However, those figures represent a minority position. When it comes to owning nature, far fewer people are landlords than tenants. While only a few receive rental income, almost everyone pays rent or a monthly mortgage or something equivalent. Hence periodic payments tell a truer story than does rental income.

VALUE EXCEEDS PRICE

Besides the matters of utility and accuracy, price has an ethical failing, too. In the modern mindset, price conveys ownership forever while rent conveys occupancy temporarily. Ownership, especially when concentrated, magnifies power while diminishing responsibilities toward community and environment.

Thinking of land as having a price, we see that part of nature as a commodity, an object of speculation. Rent—a non commodity—counters that. Periodic payments recall The Bible: "The land shall not be sold for ever, for the land is Mine; for ye are strangers and sojourners with Me."

While most specialists use "price" and "value" interchangeably, one scholar used the terms "rents" and "values" interchangeably. He was also the one guy (who led his team) to tabulate land exclusively (not land plus buildings). That was David Albouy, of U Illinois and NBER, investigating the worth of cities.

We're so used to buying and selling everything, including land, that we automatically, and mistakenly, assume that price is value; it's not. Price is only the part that the buyer pays the seller.

Buyers pay more collectors than just the seller. Value is not just price or annual rent but the entire amount one is willing, able, and required to pay. That includes:

- mortgage interest paid to banks (the "F" in "FIRE),
- insurance (the "I" in "FIRE),
- other charges the Real Estate industry manages to impose (yet may not be necessary); plus
- taxes on property.

An accurate tally of the worth of national Earth would include the land half of these four payments – rent or interest, insurance, fees, tax – in the grand total.

FLOWING TO A TOTAL

Who's going to do what the academics haven't? As usual, if you want something done right, do it yourself. Don't think me a delusional megalomaniac, able to solve grand puzzles the experts can't. Just remember:

- Astronomers helping NASA identify the landscape of the moon and Mars.
- The authors first to theorize an asteroid killing off the dinosaurs, Allan O. Kelly and Frank Dachille, and …
- Michael Faraday, who laid the groundwork for Einstein,

were all amateurs. Amateurs can go where the overly cautious conventional pros fear to go.

Experts eschew spending; we build on it. Nobody can collect from their sale of land until somebody else pays up. As rent is upstream of price, so spending is upstream of cashing in.

Fortunately for gadflies, official figures on spending for land are likely to be more accurate than official aggregates of the price of land. Political pressure influences economists and statisticians to slight land value. This pressure is off, as is the human tendency to adopt the winner's POV, when it comes to spending.

Most Americans don't buy raw land, but part of the price of what they do buy goes to rent. Most houses come with land. Food purchases pay for farmland and rangeland. Fuel pays for oil fields and mines and ores. The Bureau of Labor Statistics says in 2017 consumers spent, in trillions, on:

- shelter (inc. property tax) $1.5
- food $1.0

- utilities $0.5
- fuel $0.3
- medical care (patents) $0.6
- entertainment (copyrights) $0.4
- schooling (credential monopoly) $0.2 All from Table 2400.

TOTAL $4.5 trillion, over 1/4 of national income. If half is rent, then it's presently at least $2.3t.

> Owe, owe, owe you bought,
> Thoughtlessly, you dream,
> Merrily, merrily, merrily,
> Rent is but a stream.

Yet, the list is incomplete. It's not just households who consume but also nonprofits, businesses, and governments – the four official groups of purchasers. So, switch from purchasing to producers.

Spending underlies GDP. The US government's Bureau of Economic Analysis put GDP in 2017 at just under $20 trillion, not corrected for inflation. The GDP sectors loaded with rent occupy these portions of the total:

- F.I.R.E. (Finance, Insurance, & Real Estate) 20.9%
- agriculture, forestry, fishing, and hunting 0.9
- mining 1.7
- utilities (water, grids, etc) 1.5
- information (patents & copyrights) 4.8

All from Industry Data. TOTAL 29.8%,

Nearly 1/3 of GDP ($7t). If half is rent, then it's nearly $3.5t, over a trillion more than the $2.3t above. A trillion dollars may be small change for experts but it's like Mt. Everest for me. Yet the larger figure is reached another way, too.

Sticking with GDP, just over $20t in 2018 Q2, but now from the POV of expenses. People paid, in trillions, for:

- housing + utilities (why combine?) $2.5
- residential investment $0.6
- non-residential structures $0.6
- eat out $1.0
- eat in $0.9

- fuel $0.3

- intellectual property $0.9

All from Table 3. TOTAL $7 trillion,

Over 1/3 of GDP. If half is rent, then this way it's also $3.5t.

Yet the list is incomplete. People also pay to park and to play in parks. Consumers pay for non-solid "land" (as economists define the term); water and intangible electromagnetic waves. Furthermore, paying the property tax keeps one's title to land. Compared to other expenses, these would not be huge amounts, but they add up.

This $3.5t does not tell the whole story. Besides Earth that someone is paying for, there's land nobody is: land that's owned free and clear, farmland lying fallow, unlogged forests, capped oil wells, un-auctioned airwaves, uncompensated eco-damages, etc. Paid-off homesites alone are worth $0.7t – 1/3rd of owners of real estate priced at $28t, rentable at $2.8t. All the other natural resources together would probably go for a third trillion. Eco-losses (clean ups, health costs, regulation, etc; Ch 23) a half-trillion. Adding this $1.5 trillion brings the value to … Showtime for the greatest stat on Earth, literally: The worth of Earth in America – $5 trillion (also the global value of exchange-traded funds). This amount per-capita registered voter is $2,600/month.

That $5 trillion can be doubled. There's another class of assets that acts like nature – privilege. Like natural resources, government-granted privileges don't need anyone's labor or capital to exist. Without those two human inputs, privileges can be extremely valuable.

Privileges create rent. Holding them, business can overcharge. Licenses enrich doctors, patents enrich Big Pharma, copyrights overly endow Record companies. On and on.

The granddaddy of them all is limited liability. If you plan to put consumer and worker at risk, you can limit your liability by getting a corporate charter for a mere filing fee. Imagine if government were run like a business and charged as much as an insurance company would. That charge, the value of one privilege, could raise $1 trillion. That puts rent at $6 trillion (how much debt China keeps off its books). Per capita registered voter, that's $3,200/month.

Next, patents and copyrights. Getting one is like planting a flag on a field of knowledge, preventing everyone else from exploring there. That confers tremendous competitive advantage yet it, too, costs a mere filing fee. If government were to charge the market value of these monopolies it grants, it could rake in $2 trillion per year (Ch 24). Add that to the $6tr; now we're at $8 trillion (also the amount of global debt, all stocks in developing nations, and the tourism industry). This amount per registered voter is $4,200/month.

There's more: the money monopoly. Our Founding Fathers gave the power to create new money to Congress, yet Congress gave that away to the central bankers. That forces us to spend far more on interest, fees, and inflated prices (directly in our household budgets and indirectly as taxes) than we would otherwise spend in a competitive market. Inflation and interest on debt owed by consumers (including homebuyers), businesses, and government is well over $2 trillion, much of it rent, putting the total at $10t (minimum total in 401ks). Per registered voter: $5,200/month (the average salary for professionals).

A FIGURE FOR A NEW FUTURE

At $10 trillion, half of GDP, more than half of income or spending – nature and privilege are just as valuable as labor and capital. Yet what do economists do? Ignore half of the economy. Perhaps because privilege is a creature of politics while creation is the subject of geology, academics see both as beyond the parameters of economics.

This $10t isn't authoritative. We're not authorities. Yet authorities are not up to this task. Of all the totals by authorities, none is a measure of the money we spend for the nature we use. Ours is. Using official stats in a novel way, we showed how it's done – quit looking for an official price and start looking for an official expense. Turn from price to rent, then from income to outgo, to finally measure the worth of Earth in America.

This datum is now loose in the world, as is the knowledge of how to calculate it. Further, the method is in plain English, accessible to every lay-person. It's what one would hope for from a public agency.

Going forward, a number-cruncher would update the inputs and refine the method, maybe rely less on voluntary surveys and more on actual receipts, in order to hone in on an even more precise number.

Better than a lump-sum price, spending flows can tell you what phase of the business cycle you're in. Once you start tracking this stream, updating daily, quarterly, or as much as feasible, then you're in business as a prognosticator. Its positive changes, negative changes, they are great indicators. One can have faith in economic statistics again. Now to announce that from the mountain tops.

CHAPTER 38

THERE'S A NEW INDICATOR IN TOWN: IT'S GROUND

It is the mark of a truly intelligent person to be moved by statistics.
– George Bernard Shaw, playwright, social critic, proto-geonomist

SHOCKINGLY GRAND

You heard it here first. The worth of Earth in America is immense – far bigger than any economist or statistician previously suggested. The doubting Thomases in the economic arena may not now be believing Thomases, but at least some doubt has been sown into *their* doubt. The total does have shock value. Plus, it's supremely useful, enabling both prediction and ebullience. Tell a friend.

Some mainstream economists agree that counting rent lets one correct the figure for inflation. "The increase can be explained by increases of the price of land underlying buildings induced by money-creating, lending and borrowing, and is, i.e., inflationary." It's hard to keep a good idea down.

On the other hand, some specialists and members of the lay public may be satisfied with official figures for other economic trends yet criticize our unofficial figure for the worth of non-labor/non-capital. They exercise a double standard. What statistic is above reproof? The rate of inflation is widely regarded as flawed, the percentage of unemployed is highly politicized, GDP is exaggerated. Every quarter the BEA revises its releases, painting a less rosy picture, one that most miss, having missed the revision and seen only the original release.

When government releases its stats for GDP, inflation, and unemployment, officeholders don't rush around rewriting their taxes and subsidies. At the peak of the business cycle, businesses whip up a feeding frenzy, ignoring warning signs. Yet those official stats, unlike ours, need not run the gauntlet. Our tabulated figure is as reliable as any other grand aggregate and better than most. But what is fair when it comes to money?

REBUT BUT AGAIN

The critics who claim rent is insignificant deprive themselves of the answer to why humans foul their nest, inequality yawns cavernously, government grows obese, and the business cycle topples. All are consequences of individuals doing whatever it takes to capture rent. Rent must be immense to explain so much.

GREY OVER GREEN

Naturally, humans need to alter the environment, as do elephants and termites and many other species. Yet must that alteration be so ruinous to the health of humans and other species? Actually, no, neither pollution nor depletion is necessary. However, because humans spend so much to own or use land and resources, owners and investors do just about anything to steer that massive spending their way. And because rent, being generated by society, is something for nothing, it, to use the jargon, creates moral hazard.

The biggest sector in the GDP is F.I.R.E., the biggest in that is housing, and half of housing is land. To capture that rent stream, business wins favors for sprawl. Another large sector is energy; again, business wins favors for combustible fuels like oil that burn dirty, over incessant power sources like sunlight that operate cleanly. Further, consumers buy a lot of food, including meat; ranchers win more favors than organic gardeners, while cattle trample stream banks and emit enough methane to alter the atmosphere. At the bottom of the eco-crisis is the insider's relentless grasping for the vast flow of rent.

Counterbalancing all that, knowing Earth's true worth might make environmentalists into economic realists. It's not rational of them to let land remain such a fat profit-maker and expect a law that "just-says-no" to development to succeed. To defend an ecosystem, your planet needs profit on its side.

INCOME PARITY

Critics of inequality suspect that enormous fortunes must be unearned simply because they're enormous. However, some people are more talented, harder or smarter working, or just lucky. Yet do those factors explain the enormous gap?

What is the stuff of vast fortune and how do some capture it? Despite wanting to close the income gap, wannabe do-gooders don't know what widens it. They assume that the lever is corralling capital. Actually, it's controlling land.

"Rising housing prices are better seen as a transfer from prospective buyers to prospective sellers than a nationwide increase in wealth."
— A pair of Harvarders, Edward L. Glaeser and Joshua D. Gottlieb
in *The Wealth of Cities*

"Better seen," they say, yet almost nobody looks. What do the wealthy get? It's not wages. Many of the truly rich don't work while no group of jobs pays enough to account for the growing gap. It's not interests from capital. Capital depreciates.

What does appreciate and account for the accumulated and swelling riches is land as location plus privileges, especially patents and copyrights dished out at way below market value. Land and privilege – created by neither labor nor capital – inflate in value faster than do any goods or services offered to consumers and investors.

What keeps those geysers of wealth corralled is successful rent seeking – the labor of lobbying and the capital donated to political campaigns (Ch 12).

Bloomberg's Noah Smith, 27 March 2015: "... in order to address wealth inequality, it's important to focus on land. Even after the rise of the modern corporate economy, unequal ownership of the most basic and ancient asset of them all is still creating big divisions in our society. [I]t's landlords, not corporate overlords, who are sucking up the wealth in the economy."

STATES OBESE

No matter who's in power—whether opposed to government or not – government continues to self-aggrandize. Of course as the economy grows, so must government – air traffic requires air traffic controllers, and so on. Yet the bulk of expansion consists of bureaucracy to assist ordinary people who're needy, not receiving any shares of rent, and of expenditures on corporate welfare, to direct rent to insider elites.

Present rent recipients influence budget priorities. Rockefellers and the "oiligarachy" via their Council on Foreign Affairs define much of US foreign policy, and the military expands malignantly. Doctors receive license rent and medicare is a huge and growing expense. Bill Gates, uber enriched with patent rent, has his foundations redo classroom curriculum and global public health while ignoring alternatives that enable choice.

As income rises, and lucky recipients bid up locations, rising rent indicates ground trafficking. Some speculators keep land from best use, even vacant, precluding others from investing or working there. As the custom of property is presently construed, progress for some must mean poverty for others, as Henry George explained in *Progess and Poverty* a century and a half ago

The resultant inequality worsens health and crime and scapegoating of foreigners. Government spends ever greater amounts on addressing those two major symptoms of inequality and waging war against handy fall guys. As grows rent, so grows government.

Ironically, people opposed to government win office and control much of it. Nevertheless, even with them at the helm, government has outgrown even a mammoth-sized bathtub "to drown it in." If Grover Norquist's friends everhope to put obese states on a diet, they'd better put rents into different pockets, perhaps into the pockets of all citizens..

BUSINESS CYCLE

It's expected that economies expand and recede with the seasons, with sunspot cycle, and other natural periods. Yet, should economies chronically boom, then bust, or should they climb and glide, like an average rate of respiration?

What do postmortems on economic peaks reveal? What causes the fall in consumption and the rise in firings, bankruptcies, and foreclosures? It's not that goods or services have become unaffordable; locations have. As their prices rise, buyers spend more on that asset that nobody created and less on the goods and services that their neighbors and compatriots produce, while owners burn through equity to keep abreast.

The heady climb of site values into the stratosphere finally reaches a point where a critical mass can not afford sites. By then, gleeful developers and lenders will have overextended themselves. Spending on produced goods and services will have become too little to maintain sufficient exchange among economics actors. With customers too few and debt too massive, recession results – the gamblers' backwash terminating the mad dash for rents.

"Homebuyers will borrow the loans that we foreclose them with" is reminiscent of, "Capitalists will sell us the rope we hang them with," by Lenin. While nooses have gone out of fashion, contract traps have not.

215

Like a Grand Unified Theory, fervent and relentless rent-grabbing explains what many disparate theories cannot: e-collapse, inequality, gross government, and regular recessions. As the shiny lure which urges on so many major economic behaviors, flowing rent must be immense, and is. Measuring it is immensely useful.

THE ELEPHANT IN THE ROOM GETS A JOB

FEDERAL FORECAST

By tracking rent, forecasters can tell when society's spending on assets no one ever produced reaches a tipping point. That's when spending on produced goods and services becomes too little to maintain sufficient exchange among economic actors. Then, recession follows, regularly.

And recur recessions must, as more people bid more money for the most desirable sites. During the upswing of the business cycle – the first 14 years of this 18-year curve – land value mostly rises. Noting the rise, investors go all-in. Not knowing this cycle, many lose fortunes. If they're not a major player, government does not bail them out.

Already, some savvy firms sell accurate forecasts based on this land-price cycle – Hoyt in Florida and a couple in Australia are the ones I've found. Why don't they all? Does it have something to do with rocking the economic boat? Whatever. With a reliable figure in hand, at least we non-specialists can take advantage. Knowing what's coming lets us safeguard our savings.

UNIVERSAL SUCCESS

The point of an economy, of course, is output for people to dip into. The more goods and services flow, the more comfortably people can live. Surplus proves the perfection of production, making it a good indicator of economic health.

One definition of surplus is excess, an amount greater than necessity. Another way of looking at surplus is as a bonus, as an unexpected extra output from the applied input. A third way to regard surplus is not so much as a quantity as a quality that is, output due not to exertion—labor or capital—so much as due to happenstance, to land or location, harkening back a couple centuries to David Ricardo, economist and winner at stocks. It's the third understanding economists use when asserting land value is a surplus. Hence how much people pay for land is a surplus, too.

The more rent flows, the more successful the economy has been. Rent indicates economic health better than GDP, which ignores surplus while honing in on growth. An even better indicator is free time, a consequence of a successful economy. Leisure is society's rational use of rent. The greater the leisure, the more equitable the sharing of the surplus has been.

LOCAL WASTE

As a visit to any city shows, many landowners keep highly valued parcels vacant. What happens on vacant lots? Nothing. Other than tossing trash and occasionally shooting it up. But no one works on vacant lots. No one is investing in vacant lots.

The parts of cities where the poor go through their daily lives – that's where you see the most vacant lots. See vacant lot, see unemployment. See vacant lot, see unrealized profit. See vacant lot, see displaced development (sprawl). And sadness.

The greater the vacant lot's value, the greater the loss to society. Compare the value of vacant lots to the value of absent improvements (zero). That difference is an indicator of sad tidings. Oceans have their dead zones due to pollution. So do economies have their dead zones due to a peculiar sort of social pollution – unchecked land speculation.

SHARPENING THE INDICATOR

Gauging rent is not like measuring the speed of light (or of dark, as Steven Wright notes) or the weight of the earth or anything fixed. As long as human populations keep growing or technology keeps progressing, the worth of Earth continues to grow. Measuring economic phenomena means, like Sisyphus, you're never finished.

Although cautious "experts" lag behind the curious public, eventually they'll come around, as they always do when out-numbered and out-enthused. Discovering this datum useful, the users will want it updated, more precise. Sensitive to popular pressure (number-crunchers are people too), and wishing to mollify the serious amateur, academics will both refine the methodology and update the ongoing measure.

Then the busywork of bureaucrats would become useful tabulation. Some of the reports they issued in the past estimated a portion of rent. Those public agencies have what it takes to collect, collate, update, and publicize the size of our spending, en masse, for the parts of nature we use.

217

These agencies could make that datum a permanent part of their regular reports. That would institutionalize rent as an indicator, hopefully in a user-friendly form. Whenever the mood struck, you could find out how much your society spends on the nature it uses. You'd know your economy's bounty, and what to expect next.

The New York Stock Exchange, not far from the Fed, updates the prices of stocks once every second, and somebody does bonds. The technology and correctible rough statistics exist to track nature and privilege. Society could have the stat for all rents every day. Measuring surplus will become the new normal, and enjoy all the benefits conferred by normalcy bias. Then, of course, everybody will act like they knew about the role of Earth's worth all along.

Better than preparing for the next recession, of course, is preventing them. Similarly, better than seeing underused land's role in creating poverty is eliminating poverty. Some societies, with their eyes on this prize, redirected rents (next chapter) and took strides toward both goals.

Now that we have calculated a total for what we spend on the nature we use, the genie is out of the bottle. Is there's another shoe to drop? Are the elite down for the count? Or have we awakened a sleeping giant? Can their response shred our calculation?

CHAPTER 39

FUND COMMUNITY WITH LAND RENTS?

Since death and taxes are certain, is that why land taxes are stillborn?

REAP AS Y'ALL SOW

If you're going to have taxes, consider the advice of every major economic thinker from Adam Smith on down. They noted that a great tax base is land value. Two dozen prominent American economists, after the fall of the Iron Curtain, advised the new Russia to shift taxes off labor and capital, onto land.

Society need not even use a tax. A lease (as in port districts), a fee (such as a deed fee), dues (as in HOAs), or another fiscal tool would work as well. Historically, several places chose the tax. Having to pay it motivates owners of nature's surface to use their land efficiently. From that improvement cascades a bundle of benefits.

Given that track record of success, it makes a curious person wonder about curious governments. Why don't they measure the value of Earth in America? Since nature and privilege could combine into an awesome tax base (Ch 37), why don't officeholders have their bureaucrats figure out how much revenue is available?

Try influence. Look at Prop. 13 and its sequels. Homeowners resist giving up their land value, even when it makes their community healthier. A rational politician [sic] might worry that landowners would vote against "her/him" if they proposed public recovery of socially-generated land value. So, if officeholders are not going to recover it, why measure it?

Because. Measuring the value of sites, resources, and government-granted privileges would put a spotlight, wanted or not, on their total value – and that empowers people. Aware of that statistic, a citizen could anticipate the business cycle, to mention just one good use (Ch 38). Plus, once enough people see how much is at stake, and learn how beneficial tapping it could be, that might tempt a majority to go ahead and tap.

Just Emote "No"

I shall never use profanity except in discussing house rent and taxes…
— Mark Twain

The typical reaction to the proposal of surrendering the rental value of one's location to government is not positive. Property is a right in most countries, so paying over rent feels like a violation of that right, a negation of what it means to own a chunk of Earth. Paying the rent to one's community seems confiscatory.

People worry that it'd be harder to stay in their homes. And their homes, for gosh sakes, sit on land – their land! People live and die for land. If any policy triggers humanity's territorial imperative, that's it. All life – even ants and plants – fights for turf. Yet the fight is supposedly futile, *"For the Land is Mine"* – "God," Yet battles persist.

Other taxes on a source of rent do not trigger this emotional reaction. People are OK with taxing oil; nobody literally lives on oil. And they're largely indifferent to higher fees for patents – which unbeknownst to many could rake in trillions annually – since tech giants could easily afford to cover their claims on the field of knowledge. But for us land animals, taxing territory is a totally different matter.

The tax collector (government) is the institution with unmatched power. In an autocracy, they could charge as much as they like, evict anyone in their way. Their demanding rent would turn government into a land over-lord. Or so some worry. And not without reason, especially if one has had dealings with the IRS or felt pushed around by government exercising eminent domain.

Some angry opponents label a tax on land as ultra-leftist. However, if akin to any ideology it'd be Christian, with its notion of surrendering to Caesar what is Caesar's, or, pay over to society the values society has generated. Further, between the world wars in Eastern Europe, when the proposal of taxing land was gaining popularity, it was a coalition of leftists (unionists) and rightists (gentry) who defeated basic forms of land reform.

Returning to the powerful human attachment to territory … It is not fixed in strength nor in definition; it has changed over time. Now we think of this attachment as individual ownership. However, up until modern times, it was for communal ownership. Land was not "mine," but "ours."

As commonplace as our practice of buying land may be today, earlier humans saw the notion as absurd. When Lewis And Clark crossed the

Louisiana Purchase and met the hunters-gatherers already settled there, they had to explain where they'd come from. They'd cite:

- houses that floated (ships), and of course the tribes doubted that;

- cities packed with more people than all the surrounding tribes counted together, and the natives scoffed, "Yeah, right." Then Lewis and Clark claimed that ...

- individuals owned the land. The Indians burst out laughing and said that now they knew the white man was lying all along, because nobody could own the land.

Virtually every culture – if you go back far enough – made the argument that "nobody can own the land" or "the earth belongs to us all," etc.

However, once tribes settled down to farm, then communities did own the land. Next, when community settlements increased and families ruled, the head of the family owned the land. Later, as population and independence grew, yes, an individual could own what was once Mother Earth. Now, with thick density and cold autonomy, individual ownership is the new normal.

In their gut, our pre-agrarian ancestors knew that paying any one person for land was irrational and wrong. We moderns don't know the feeling. We feel the same way when paying for land as we do when paying for anything else. We feel the same being paid for land as being paid for our labor.

I'll bet when slavery was customary, people felt no differently paying for a slave than they did paying for a stave (to whack the slave with). Or being paid for selling a slave or a stave. Yet, hardly an honest soul today could even conceive of profiting from human trafficking.

MORALITY OR NORMALCY BIAS?

Ironically, an objective study of the issue shows these fears to be unfounded. What should be feared is government's *failure* to recover socially generated values. Where governments thus fail, elites gobble up the rent. They erect a steep hierarchy, widen the gap between haves and have-nots, and behave more oppressively. The World Bank showed this true for nations, but it applies to states, too.

While some voice outrage about paying land value to government and having government confiscate their land – things that remain largely figments of overheated imagination – their outrage is muted about paying mortgages inflated by speculation and banks foreclosing on homeowners; actual tragedies afflicting sizable numbers. It's banks, the property of elites, who do the most confiscating of land; governments not so much.

Governments do confiscate a portion of income directly and indirectly, by taxing sales. Most people consider such taxation as moral or amoral but not immoral, despite those values being individually generated as opposed to socially generated, as are site values. For most of us, socially generated value is a foreign concept.

What people do complain about is the amount. A land tax would be yet another tax. While true, how much people would pay equals how much mortgage people would not pay. That's so because as the tax on land goes up, the price of land goes down. When society chooses to "socialize" land rent, none is leftover for speculators and other owners to *capitalize* into price. Thereby the land tax does not change how much one pays but to whom one pays it, to neighbors rather than to a banker. That keeps it in the family, so to speak.

Despite paying over more land value than they'd like, people do receive a bigger bang for the buck. While it's a small sample, the State of New Hampshire only levies a property tax – falling on both land and buildings. The Granite State ranks much higher than most other states by almost every measure.

More irony: the only way to solve foreclosures is to levy land. When owners must pay out site rent rather than keep it, speculators lose their reason to withhold sites from use. They reverse course and develop their sites. The resultant greater supply of sites in use also lowers what buyers and tenants pay for underlying land. So, mortgages shrivel and with it a buyer's inability to pay. Thus foreclosures, too, decrease.

Turn from emotion to cool self-interest. Most people own some land themselves (whether outright or not). Many view this payment as a tax on the middle class, and as a tax that would spare the rich. Which is ironic; it's an urban myth the rich are happy to hear. They know where their fortunes come from – downtown locations. Their parts of cities, New York, London, Tokyo, et al, have the spendiest real estate on earth. They also corral the rents paid for oil and privileges like monopolies on fields of knowledge (patents and copyrights). A tax on land would mean far heftier payments by them than by anyone in the middle class.

Furthermore, sad to say, many Americans have no more savings than the equity in their – not homes, as we're want to say – but in their location. Stack all these objections together. Of course, the property tax – close cousin of a land tax – is the tax most people most love to hate.

MOTIVATED OWNERS UP THE OUTPUT

If a policy is beneficial, does that mean it's also ethical?

Opponents to the idea of taxing land claim it'd discourage development. Yet the reverse is true. Where government fails to recover ground rent, a few owners can own more than they can possibly use and keep many prime sites out of productive use. Hence the economy is sluggish, both investment and wages are low, the income and wealth gaps are wide, and poverty appears intractable.

To correct that, advocates of geonomic logic as laid out by Henry George, the follower of Ricardo who in the 1880s was the third most popular figure in America after Thomas Edison and Mark Twain (one of George's admirers), found legislators willing to listen. Hence, some jurisdictions have levied land. Most shifted their property tax off of buildings, onto locations.

The reform delivered the desired results. The notion that the worth of Earth could benefit everybody is not pure theory. It is also sound practice.

Having to pay land dues spurs owners to put sites to good use or sell to someone who will. Places that tax land, not buildings, attract investors. Honolulu developed its Waikiki Beach, and Pittsburgh renewed itself totally from private investment without one penny of public subsidy. The Steel City was named America's Most Livable City two years running.

Like the old carrot and stick, the levy on land spurs owners to develop vacant lots and parking lots, while the zero tax on buildings entices owners to improve or replace antiquated or abandoned buildings. Those owners erecting buildings infill cities. In compact cities, residents do not rely on cars so much and walk more, ride bikes and take buses.

Happily for life forms, tax-spurred development is tempered, benefitting the environment. Where the community does require owners to pay "land dues" (or the equivalent), owners tend to take no more than they can use and to use that wisely. Those kinds of decisions reduce both depletion and pollution. Without even thinking about it, people leave behind fewer byproducts that pollute (Ch 29).

Owners engaged in development add to the housing stock; as supply goes up, the price of housing goes down. Plus, paying the rental value of locations as land dues removes that value from the price of housing; locations of reduced price mean housing of reduced price, too. During the two decades that Pittsburgh shifted its tax from buildings to land, their housing was prize-winning affordable.

With our current relationship to land as property – versus land as "trusterty" in the language of Ralph Borsodi – we guarantee ourselves periodic recessions. When bankruptcies mount and the newly unemployed line up at (un)employment offices, our recurring recessions slash the ranks of property owners. Our self-imposed recessions keep ownership a dream for many and a memory for others.

While property is wonderful, it's even more wonderful when more people can hang on to it. Which they can, when their economy is not in recession. Any place can avoid a downturn. Certain Australian towns that taxed land opened new factories ... during a recession! In neighboring towns, factories were closing. If other jurisdictions were to duplicate what smart towns do, drastic recessions would become a thing of the past.

When owners develop vacant lots and rehab abandoned buildings to earn more money in order to pay their land dues, they develop the entire region, spur the economy, and spread prosperity. Constructing all the new high-rises, etc, and running businesses out of them, employs people. New Zealand knew 99% employment for an entire decade.

Meanwhile, Denmark got inflation down to 1%. What the Federal Reserve could not do – fulfill its mandate of solving unemployment and inflation – geonomists in various places did do. Of course they had to follow reason and confront political reality, yet in doing so they produced a more impressive resume than most other reform attempts.[1]

Singapore, along with Hong Kong, is often rated as the best city in the world for doing business; plus, the freest city. The people doing the ratings are capitalists with their own ideas of freedom and justice. Those raters note both cities levy fewer and smaller taxes yet those judges leave out the fact that both places recover lots of socially-generated land value.

In Singapore, all the land is private, owned by private citizens, but taxed at a high rate. In Hong Kong, all the land is owned by the public and leased to the owners of buildings. Semantically, there's a difference, but functionally the two policies are pretty much the same. And the benefits are laid out for all to see: prosperity, big middle classes (now shrinking in places like America and Germany), and enough public revenue to afford quality public services, and even a dividend, in Singapore.[2]

For many people, it's hard to believe that public recovery of site value really works that well. And if it's such a good idea, why is not better known and more widely used? All these examples are from the past, not the pres-

1 "Successfull examples of land value tax reforms" at P2P Foundation, 5 Feb 2011
2 "Singapore to pay bonus to all adult citizens after budget surplus" by Yvonne Rarieya at CGTN, 20 February 2018

ent. Even when society took these first steps toward economic perfection, they immediately got derailed. Repeatedly. Why? How?

Public recovery of social surplus sows the seeds of its own destruction. It gets repealed because it works. It greatly improves an economy's performance. Its participants, getting more income, spend more money on land, pushing up location value. Speculators see that climb in price and want it for themselves. Given the political power of the real estate lobby, they get the landward tax-shift repealed much more quickly than it ever took proponents to get it adopted.

THE POT STIRS

Where Americans fear to tread, others move forward. More confident Chinese, unlike rulers elsewhere, are better able to face reality, more willing to try what works. They buy prime land worldwide and move to tax property at home which contains a land tax.

On the investment side, the Chinese now own princely properties in America, as did the Japanese before them back in the 80s, and the Germans before them, and the English nobility forever. Investors from abroad are not under the thumb of the American rentiers. Rather than take "unknown" or "irrelevant" as an answer for the worth of the Earth in America, I'll bet they find out the true value of what they purchase. To get their money's worth, they may also calculate the phase of the land-price cycle.

What's different about the Chinese compared to other foreign investors, though, is they're bigger, faster growing, more rational (centralized), and have trillions of dollars to spend. If they buy up great parts of America and that worries you, don't let it. Just recover land value and enjoy the benefits. Plus, your public treasury will be flush.

If the worth-of-earth figure were available and were huge and if you're not getting any or much of it, then you should know it's concentrated. Not that you'd do anything about that. But the few who now get the lion's share, they don't know how obedient and lethargic you are. They might worry that a fat rent figure could light a fire under you, and make bounty a boon for everybody, not for just those lucky few. Social movements need such fires.

Without even knowing the grand total of the value of all land and privileges, some bold Americans do promote the public recovery of location value. Economic nerds like the policy because it spurs efficient use of land.

- Many commentators abroad; here: Slate, Sightline, *et al.*

- At *Fortune* magazine, a blogger, Tim Worstall, shines a positive light on the land value tax for his readers.

- At Bloomberg – a billionaire's website for businesses – a columnist, Noah Smith (no relation, surprisingly), calls for a real solution, no matter how cutting edge and ahead of the pack – tax the value of locations" (9 September 2015). He adds, for affordable housing … *"there's one very powerful policy that cities and the activists who love them haven't yet employed -- the land value tax."* A land-value tax is an efficient and fair way to take a city that now works only for lucky prosperous landowners, and turn it into a place where the working class can afford to make a decent life (24 October 2017).

Those voices, plus knowing the size of land value as a potential tax base, could give timid politicians the confidence to take a page from the Pittsburgh playbook and shift taxes off of people's efforts (wages, sales, and buildings) and onto the low-hanging fruit produced by nobody – the ground beneath your feet.

To deal with possible losers, if any landowner could show a net loss after paying land dues, then the government could pay such deprived owners a bond. A few unlucky souls should not have to pay the cost of a reform that benefits the many. Governments would not have to pay a lump sum but an annuity out of the recovered rents.

Beyond just net losers, government could pay everybody; combine the taking with a sharing. Paying a literal dividend or a subsidy to all residents is how Aspen won its sort of land tax, how Alaska passed its oil tax, and how British Columbia passed its carbon tax. For most residents owning land of average or less value, the dividend would exceed their land dues, since commercial sites pull the average up so high. Money in the pocket – a rent share greater than the land dues one pays – makes the paying in more than palatable. It's almost a no-brainer (next chapter).

CHAPTER 40

Surplus – A Way to Remake the World?

Misers make wonderful ancestors.

Dividends for Everyone?

Lord Bertrand Russell, who lived to almost 98, was asked, *"What's the difference between ignorance and apathy?"* He answered, *"I don't know – and I don't care."* After attending Cambridge, he co-authored *Principia Mathematica*, which in 900 pages gave math its logical underpinnings. Russell also proposed that society share natural rent. Is that what we should do with the immense ocean that is the full worth of Earth in America?

Take another peek at how much we spend on the nature we use (Ch 37). That hefty flow of bounty challenges our chronic assumption of scarcity. No longer is the economic challenge we face production; now it is distribution.

So how should we distribute this torrent of socially generated values? There are four different ways to divvy up that pie. Should we:

1. Leave it in the deep pockets of those now capturing the lion's share?

2. Put it all, via taxes, into the hands of politicians and bureaucrats to spend as they see fit?

3. Shrink it to insignificance via land trusts and price controls? Or ...

4. Expand the Alaska model nationwide? That is, share it.

My formula for success is rise early, work late, and strike oil.
– J. Paul Getty

Handing over rent to government (#2), is not popular. Shrinking its regulated amount (#3), would spawn black markets. Doing nothing (#1), means we're OK with the current capture of rent (half of the GDP).

Or take a page from the Aspen playbook and turn the value of locations and natural resources into a source of funding for something like Social Security for everyone (#4). Then nearly everyone would take an interest in measuring the size of society's surplus. Conversely, were government or academia to calculate the worth of Earth in America, that figure would likely intrigue people with the question of what to do with so much bounty. No matter what we choose, once we know its size then we know the health of our economy and when to anticipate its ups and downs.

EMOTIONAL ANTIS –

Just as many of us feel emotionally predisposed to reject paying a tax on land, so do many of us feel the same way toward receiving a share of rent. Even though it'd mean fresh money in our own shallow pockets, many of us oppose getting something for nothing ... in general. But not absolutely.

While some object to public spending on the poor, they do not object to politicians providing corporate welfare to the rich, which pays off big-time. A sizable portion of enormous fortunes are not earned. Didier Jacobs, senior economist for Oxfam America, estimated in 2016 that about 75% of U.S. billionaire wealth is derived from rents.

> ... "rents" is a word that means "leveraging control over something that already exists, such as land, knowledge, or money, to increase your wealth" ...
> – Rex Nutting, *MarketWatch* of CBS

Those who admire the unduly fortunate and oppose largesse for the innocent unlucky extend that double standard to embrace themselves. They do not clamor to repeal the handouts they get: Social Security, COLA, Medicare, public schools funded in part by childless homeowners, etc. For them, a freebie is good for the goose, not the gander.

Further, objectors claim many recipients would waste their share, even though they do not object to current rentiers wasting more than their share on a profligate lifestyle. Yet most welfare recipients don't; those poor who receive cash, no strings attached, actually spend the money in socially approved ways. As many poor say, they want a hand up, not a hand-out.

Critics fear many recipients would quit working despite the fact many can't keep working because their jobs are disappearing. It's good riddance in many cases, since a huge number of jobs are not productive but merely conformist; the work does not actually produce food, clothing, shelter, energy, transportation, medicine, or recreation. Indeed, most work is a waste of life.

If objectors were to look into right and wrong more deeply...

Morality & Norms, Old & New

Most of us believe that owning land means owning its value. We feel no moral imperative of owing its annual value to our neighbors. Yet by occupying land, one does displace all others. Is that fair? Not according to some sage voices.

Most moral traditions make statements like the one in the Bible: *"The fruits of Earth belong to everyone."* The phrase "commonwealth" used to be commonplace. Why?

First, nobody made Earth, so our spending for land and resources never rewards anybody's labor or capital. None of us can buy land from the maker. You buy goods and services from their producers and providers, but not land and other natural resources. That's why spending is of two types, rewarding either effort or status.

Second, enjoying an equal right to life, and needing land in order to live, we all enjoy an equal right to land. It's a right that occasionally conflicts with instinct, as when long-term residents oppose short-termers – "You don't belong here." Belong or not, within our group, when we claim or occupy a location, we displace everyone else. Hence, we owe them compensation, just as everyone else owes us compensation for displacing us.

Third, while nobody made land, everybody makes its locations valuable. It's not owners but the presence of society that generates land value. It's not old-timers selling out and moving on but newcomers buying up and settling in. The most powerful and accurate generator of land value is population density, something no owner by himself can claim responsibility for. Thus, our popular interpretation of property is a shared delusion.

Our paying dues, then getting dividends, differs from our paying sellers to leave the lot behind. Instead, owners pay their neighbors in order to stay. It's compensation for having displaced them. That feels different from the indifference one gets when paying any other price. It feels more like contributing one's share to a common kitty.

These rents we pay for non-produced land is a surplus. What happens to a surplus? It triggers the paying of dividends to the deserving; here, the members of society. Since the populace conjures rent, sharing it completes the cycle – *incipients* of rent would become recipients of rent.

229

Were modern humans persuaded that sharing the gains from land is just, they'd not be the first – far from it. For millennia sharing was customary. Gatherer/hunters shared their pickings and game. Recall the Indians in the Pacific Northwest with so much natural wealth (even if primitive); they lost their material insecurity. One did not show off and gain status by hoarding stuff but by giving it away. Their potlatches still exist today.

At the beginning of the Agrarian age, the whole harvest went into the community granary and back out in equitable shares. Later, kings were supposed to disburse this output equitably. Of course, royalty got away from that duty, and just kept the lion's share. One exception was early in the Industrial Revolution when England paid people a stipend to buy bread.

People are basically generous. In the past, workers via their unions did their best to take care of the poor. Presently people are still generous, donating to churches and charities. Even those who weren't too generous on their way up became charitable once they arrived at the top. It's in our genes.

Basic Income Grant: BIG

With power comes responsibility.
–French National Convention during their
revolution in 1793; they also funded France
with a land tax until war became too expensive.

Lately some big names who've made it big materially propose paying an extra income to everyone – a "Basic Income" (enough so people could at least scrape by) – whether they have a job or not, whether they're wealthy or not.

Those luminaries are the latest to advocate an extra income. The idea rears its head every generation:

- during the Great Depression in California, Sinclair Lewis with his EPIC,

- later last century, Robert Heinlein, the Futurists, and Buckminster Fuller,

- in Canada, proponents called it Social Credit; now ... Everywhere it's entrepreneurs.

Big names in Silicon Valley promote a Basic Income as an antidote to job loss:

- Bill Gates, Microsoft co-founder,

- Elon Musk, SpaceX CEO,

- Mark Zuckerberg, Facebook CEO,

- Chris Hughes, co-founder of Facebook,

- Stewart Butterfield, Slack CEO/co-founder and co-founder of Flickr, and ...

- Sam Altman, Y Combinator & president (with Airbnb and Dropbox in its portfolio).

Maybe they figure what's good enough for them – freebies from government – is good enough for everybody.

Also climbing aboard this bandwagon are:

- Stephen Hawking, author, speaker, astrophysicist (deceased),

- Richard Branson, founder of Virgin, and ...

- David Simon, creator of the popular HBO series "The Wire," and most recently "The Deuce."

Organizations with various perspectives endorse the idea:

- the International Monetary Fund (IMF),

- the American Enterprise Institute, a conservative think-tank, and ...

- FastCompany, a news website.

Politically:

- The State of Hawaii passed a resolution in favor.

- In Congress, Rep. of Minnesota Keith Ellison expressed his support.

- Several governments planned a trial run for a limited number of recipients in 2017. Stockton, California planned to test a basic income. At Cash Conference in San Francisco, the town's mayor, Michael Tubbs, said the program isn't a response to encroaching technology. "Basic income isn't about a scary future where robots run everything. It's about today, when working people can't afford rent." Note that word, "rent."

Canada, Finland, and Kenya have started paying randomly selected citizens. Finland has inspired Scotland to follow suit. Kenyans invested their windfall in home repairs and schooling their kids, plus domestic violence has fallen.

As for funding Basic Income, proponents have yet to find consensus around its source. None of those proponents appeared cognizant of social surplus. If the source of extra income is anything but surplus, then when recipients spend it, they'd inflate prices. Landlords in particular would just raise their charge. Note how:

- Social Security increases pump up rents in trailer parks,

- Fatter student loans pump up rents in campus communities, and

- More generous medicare gets soaked up by nursing homes.

Yet an extra income for all need not inflate prices. That just depends on its source.

RENT SHARE

Rather than Basic Income, the extra income could be a share of surplus.

How would that work? All of us who are owners would pay amounts commensurate with the value of the locations we claim and all of us who are residents in the region would get back amounts equal to how much everybody else gets. Land dues in, rent shares out.

Using fees or dues or taxes, government would redirect society's spending for land, natural resources, the electromagnetic spectrum, ecosystem services, and the rest of the planet that we find useful enough to be willing to pay for it, etc., temporarily into the public treasury then back out. Call it a Citizens Dividend.

The policy of disbursing some of the worth of Earth does have a pedigree – endorsements from some world-class, open-minded thinkers:

- in the 1700s, the physiocrats which included Thomas Jefferson and Paine;

- in the 1800s, John Stuart Mill in the UK and Henry George in the US, and ...

- in the 1900s, Lord Bertrand Russell and George Bernard Shaw, who once quipped, *"Youth is wasted on the young"*;

- in the 1960s, Martin Luther King, while citing Henry George, noted how economies constantly put out growing amounts of surplus, and proposed sharing it. (MLK also made popular the century-old adage of crusaders for social reform: *"The arc of the universe is long, but it bends toward justice."*)

How big a dividend might a citizen receive? Initially, due to politics and normal people's resistance to change, it might not be that much – maybe $100 per month. To win even that much, recall that the biggest household budget expense is land beneath housing (Ch 20). A politically palatable first step may be to replace the property tax with a land-use fee and use that revenue only for a resident's dividend, *a la* Aspen.

Eventually society would tap all rent streams. The per capita amount comes to $10,000 per month (Ch 37). Even if that calculation is off by a factor of 10, wouldn't it be nice to get $1,000 each month?

When the economy cranks, swelling land values and rent shares, one can labor less. When the economy coasts, shriveling land values and rent shares, one can work more. The economy becomes balanced, no longer needing to relentlessly grow. In both its phases (cranking and coasting), one will have enough. That solves the classic economic problem – suffering from poverty.

WELL-BEING

More money in the pocket means that one can afford medical care while not needing much medical care. Presently, medical care is the fastest growing expense in the economy. Much of the illness we endure is due in part to anxiety, whereas receiving a Citizens Dividend eases one's money worries.

How bad is it? Among men between the ages of 20 and 35, the leading cause of death is suicide. Anxiety and depression were predicted to be the second biggest causes of ill health in Western countries by 2020.

Conversely, when people feel more secure, they feel less need to self-medicate, weakening their health. When people in underdeveloped nations (including Honduras, Nicaragua, and Tanzania) receive free cash, they consume less tobacco and alcohol, showed World Bank researchers David Evans and Anna Popova. It should hold true in the developed world, too, I imagine.

Having a home is to have health, not just because dining in beats eating out. Also, owning within a community of mutual identity creates peace of mind. With a rent share, residents can afford to live where they love, even as neighborhood sites rise in value. Perhaps such payoff could ease any tension between newcomers and old stayers.

233

Another major stressor is the degraded environment. All those toxins not only poison humans and other animals but also, as the health of ecosystems fails, the worth of Earth falls. That in turn would reduce the dividend. Since most would prefer a greater dividend, they'd minimize their footprint and support policies that discourage pollution and depletion, thereby maximizing land value. Self-interest would align with eco-interest.

If residents get a slice of natural value, then owners and local developers would be compensated for not exceeding the carrying capacity of their land. Owners could profit from building or otherwise using their land, but its rental value would go into the common kitty and back out again as the dividend. By keeping their land healthy, residents would increase the dividend for themselves and for their neighbors, too.

One more way we hamper our health is via overwork. Employees working more suffer more and so produces less. Conversely, a shorter workweek does increase morale and health, not to mention productivity – and more from less is the whole point of an economy.

Already 40 hours per week exceeds our needs. How short could the workweek be?

- Marshal Sahlin figured that our Stone Age ancestors' full-time is our half-time.

- After the Black Plague left far fewer people for so much cultivatable land, James Rogers figured that in the Middle Ages our peasant ancestors actually worked 14 hours a week.

- Bucky Fuller figured two hours was enough for actual goods and services.

- Juliet Schor, while at Harvard, figured it'd be 6 1/2 hours by 2002 if productivity gains did not go overwhelmingly to the 1%.

As long as the rent share is hefty enough, paying in would pay off. With that monetary cushion, we'd make the economy work for us instead of we for it. We could shorten our workweek and improve our quality of life.

Just as jobs sicken, time-off heals. You can hang out with friends and neighbors; maybe even have dinner together as a family. Go spend time in nature, spend time defending nature. You can play more, develop new skills, realize lifelong ambitions, and lead a more fulfilling life. You'd feel good. Paying land dues upfront is like investing in a more comfortable lifestyle for life. The Protestant Work Ethic becomes an anachronism as the Polynesian Play Ethic becomes relevant.

> *Lack of money is the root of all evil.*
> – George Bernard Shaw

Turning from individual health to societal health, as prosperity spreads, the motive to steal becomes less common, as it is in Northern Europe. Furthermore, materially secure people become more hesitant to go off to war. And that's without winning any international agreements or signing any treaties between erstwhile belligerents.

What if, in the Mideast, the high value of locations in Israel were shared with both richer Israelis and more numerous Palestinians, while Palestinian markets were opened to Israeli businesses? That'd establish some cross-border community, right? This practice of sharing rent could deliver both peace and prosperity in our time.

Redirecting rent solves the problems that concentrating rent creates. Sharing natural bounty, thereby creating community, lets us unite our views of nature and real estate, balance work and play, streamline swollen governments and big businesses, and de-motivate our senseless mistreatment of one another. And, it improves economies.

PRICES PRECISE, CHOICES EFFICIENT

With those concrete goods is an abstract one – precise prices. Prices guide and misguide our economic choices. For choices to be wise, prices must be accurate.

Prices would not inflate when citizens spend rent dividends. Public recovery of site value, via land dues or land tax or land use fee or land lease, would keep landowners in competition among themselves. If any of them tried to raise what they charge, their prospective buyer could look elsewhere and their tenant could move to a new building that another owner erected, in order to have income to pay the land dues.

Rather than distort prices, sharing rent would correct prices.

Take land. Speculators constrain its supply while people crowding into cities exaggerate its demand. But with land dues, we motivate owners to use land efficiently and with rent dividends we enable people to move to human-scale cities. Improving supply while modifying demand means the price of land becomes accurate, and our choices influenced by that information become rational.

Take labor. Currently people chase jobs instead of jobs chasing people, what with jobs being taxed and deprived of some prime locations. So labor is underpaid. But when everyone, not just the rich and poor, gets money without working – a rent share – people become less desperate for jobs; they can be picky. To persuade people, employers would raise the wages they offer, improve work conditions, and tolerate flexible schedules. Work could even become something most people look forward to.

Bosses who've not thought this through can be passionately opposed to paying higher wages. Others see the upside. Henry Ford did, when he raised his employees' wages so they could afford to buy his cars. With Citizens Dividends, that phenomenon would occur society wide.

Along with stagnant wages, some worry about disappearing jobs. It's not news; new technology always eliminates old professions. Where did all the blacksmiths go? However, the same techno-progress obviating jobs also pumps up site values in desirable areas; enriched inventors and investors who, besides earning profit, also receive corporate welfare, bid up the value of land (see Silicon Valley). By recovering socially generated site values, society would be harnessing modern automation for a comfortable living for all.

Take capital in the sense of tools, machinery, factories. With extra income, inventive people could spend less time in non-creative jobs and more time inventing. Contributing their ideas and breakthroughs would accelerate techno-progress, letting us get more from less, cutting the cost of living. As prices deflate – just the opposite of ruinous inflation – physical capital would only need a steady income to make profit.

Take capital in the sense of mounds of money; fiscal capital. Currently, without a share of rent, people borrow more: mortgages, car loans, student loans, credit card debt, etc., and the middle class shrinks. Conversely, getting extra income, consumers could avoid taking on so much debt. Less demand leads lenders to lower their rates.

Businesses, too, could borrow less. Their unburdened employees produce more, and their endowed customers purchase more. A booming business not only provides more working capital but that harmonious scenario also attracts investors – an infusion better than borrowing. Again, bankers would need to cheapen their loans as prosperity expands.

Regardless of who wins the election they have to raise taxes to pay for the damage.

Governments, too, could avoid taking on so much debt. Even now governments could already cut things like corporate welfare without harming anyone.

Then later, when people are prospering, the citizenry would not need quite so much in the way of social programs. Politicians could reduce them. Materially secure people grow less warlike; we could cut military budgets, too. As solvent governments borrow less, once more bankers must lower what they charge.

Presently, to finance excess debt, the Federal Reserve creates excess new money (for the borrowers), excessive of new output; that imbalance inflates prices. Conversely, when governments, businesses, and consumers borrow little and only existing money that has been saved, that balance stabilizes prices. Minimizing debt in society translates to less inflation (distortion) of prices. With better (undistorted) information, people make better choices.

Furthermore, as labor-saving devices advance—accelerated by inventors and everyone receiving a Citizens Dividend—production costs would shrivel. The cost of living would follow—the dreaded deflation. The supply of money needed to circulate would shrink. Instead of being a printing press, banks would have to become furnaces to eliminate old money. If they can't bring themselves to do it, the public treasuries could.

Who receives the flow of rent and what they do to get it impacts our wellbeing and shapes output's prices. That in turn determines whether our economy will be effective and fair, our ecosystem healthy, our democracy sustainable, and our society humane. How we decide to distribute the surplus matters.

Society Shifting toward Sharing?

Given human nature and being used to the status quo, normalcy bias tells us this sharing of land value sounds too good to be true. Yet true it is.

Alaska pays all its residents some of the surplus it generates from the sales of its oil. The US talked about having Iraq do the same, but other voices prevailed. Elsewhere in oil-rich countries, reformers call for adoption of the oil dividend.

Every man is guilty of all the good he did not do.
 – Voltaire, who also said the
fruits of the earth belong to all of us.

Inspired by Alaska, nearby British Columbia pays residents a tax credit from its carbon tax.

Most places don't have oil in the ground but every region has a sizable downtown, and that's where the money is, even more so than buried under desert sands or Arctic tundra. That's the goldmine: urban location values. Just ask F.I.R.E..

Once the public sees the size of natural values, many see that immense figure as a windfall and feel comfortable with their community recovering this socially generated value. Overcoming our widespread attachment to "property" value, some places do recover some socially generated location value and disburse it to residents. Rather than fret over newcomers with deep pockets making the old neighborhood no longer affordable, housing advocates have treated swelling land values as a windfall.

In Aspen, Colorado, where vacant lots sell for well over $10 million dollars, the police, merchants, and even realtors could no longer afford to live there. The wealthy in that ritzy, mountain-high ski resort were not too pleased by the loss of their waiters. So, without encountering too much opposition, the local government sliced off a tiny sliver of that spending on land – the greatest portion of the price tag for housing – and used the revenue to help working families stay in town. Even doctors earning six figures qualified for housing assistance.[1]

Besides Aspen, Hong Kong and Singapore tap land values for public revenue, spending much to house the populace and meet other social needs. The Asian cities also keep counterproductive taxes low. And Singapore even pays a dividend.

In the not too distant future, other places may catch up. The tech billionaires pushing Basic Income draw attention to the concept of an extra income for all in general. The rich entrepreneurs who realize that Basic Income inflates prices may refine their proposal to base it on deflationary rent.

American society is approaching or is already at a tipping point where a critical mass questions work and the traditional workweek. Some young adults especially, many tens of thousands of dollars in debt for student loans, have their doubts about rote work. They envision an end to meaningless work, even to an end of the financial definition of useful contributions.

It's usually the youth who make change. And if not this generation, then perhaps the next. Those children naturally share resources; they think it's fair.

If you think the gain from sharing surplus is impressive, just wait. Sharing rent amplifies the gains from sharing rent. Now that we have an idea of the size of surplus, we can extrapolate how much bigger it'll be later, after we divvy it up equitably. Once bounty flows naturally, we can lock the hood on the economy.

---

1   "In Aspen, even doctors need affordable housing" by Nancy Watzman in the *Colorado Independent*, 1 June 2015

CHAPTER 41

$10K PER MONTH, FREE – IS THAT INSANE?

Money can't buy everything but what it can buy is worth every penny.

SPEND RENT, GROW RENT

Ready to go where no one else has gone before? As did Einstein, who went alone into quantum physics? Picasso, who went alone into abstract art? Steve Jobs, who went alone into the personal computer? While we may not be in their league, the payoff is. And nobody but us is about to calculate the coming worth of Earth in America. We'll just follow logic to wherever it may lead.

If you thought half of GDP – $10+ trillion – were a lot of rent, that's just how much it is now when so much is concentrated in the deep pockets of owners, sellers, and lenders. How much would be the value of land and privileges if it were shared equitably? More? Less? The same? So much that at least it'd make half of GDP seem credible today?

Actually … brace yourself for more good news. Just as taxing land value increases the tax base, so does sharing recovered rent increase the "dividen base" (to coin a term). Material security coupled with an efficient economy lets us shift spending away from negatives like fighting crime and healing from pollution into positives, like making cities livable. Within a number of years, it looks like social surplus could reach the Pareto Optimum. I know, crazy, right?

SEVEN WAYS RENT GROWS

In *The Sorcerer's Apprentice*, young Mickey Mouse could not stop the magic brooms from profligating (to coin another term).

If owners paid land dues to compensate for displacing others, and residents received rent shares as compensation for being displaced, that would level the economic playing field. On it, members of society would make decisions that benefit not just themselves but their fellows, too. And as with Adam Smith's rising tide, it'd lift all rents.

Here's how Earth's worth would grow when we all get a share of rents. Right off the bat we will ...

- (1) reduce crime,
- (2) redevelop metro areas, and
- (3) improve mobility.

Those three steps forward entice and enable residents and businesses to spend more for locations, swelling land value.

- Next, within a short while, we'd likely
- (4) reform public revenue.
- Government would quit rewarding polluters. Citizens would be physically well, while getting dividends makes them mentally well,
- (5) slashing medical expenses.

Those five advances have two major consequences for the economy, and

- (6) the GDP grows, and
- (7) the cost-of-living shrinks. Together, these seven main factors will make the size of rent, tomorrow, dwarf its size today.

How much more rent would each of these seven factors fetch? Take a look.

I. Safe Neighborhoods

Crime is costly. Victims have to make up for losses. To protect and serve, governments invest heavily in police forces.

Further, crime is distasteful, viscerally. People pay more to live in safe neighborhoods, less to live in crime-ridden neighborhoods (duh). But how much less? And how much more?

In 2012, the Center for American Progress calculated the direct annual costs of violent crime in eight cities – Seattle, Milwaukee, Houston, Dallas, Boston, Philadelphia, Chicago, and Jacksonville. It totaled $3.7 billion per year, an average of $320 per person per year. A 10% reduction in homicides would increase "housing" values 0.83% the following year[1] (actually, home sites).

Though number-crunchers say "housing," they should say "location." Three German scholars agree, saying in the portal of the Centre for Eco-

---

1   "The impact of crime on property values" by Martin Maximino at *Journalist's Resource,* 12 March 2014

nomic Policy Research, *"rising land prices hold the key to understanding the upward trend in global house prices."* It's not buildings but neighborhoods that become livable, thus valuable.

And there are more kinds of crime than homicides. Plus, the rate can fall by more than 10%. So, site value can rise by more than 0.83%. Indeed, check out what these researchers found: *"Zip codes in the top decile in terms of crime reduction saw property value increases of 7–19% during the 1990s."*

Besides the loss of site value, people have to spend extra on repairs, replacements, healing, etc, and governments on cops, courts, jails, etc. *"The costs of crime in developed countries might be 10% of the GDP or more (Entorf and Spengler, 2002: 91), which is consistent with estimates that the costs of crime in the United States might be around $1 to $2 trillion per year (Anderson, 1999; Ludwig, 2006)."* (Of course, the impact of all violence on GDP worldwide – individual [crime] and societal [war] – is greater, over 13%.)

Much crime is simply a short cut for stuff and for status. When people get stuff and status by legit means, they feel good about themselves and treat their neighbors decently. Most people are as honest as they think they can afford to be.

That is, secure people commit less crime.

Once getting a share of rent, for being a proud member of one's community, former thieves would largely quit misbehaving. Their neighbors no longer would have to make up for dead losses. Governments could exert less force and spend fewer public dollars on law and order. Lawmakers could cut taxes or kick back revenue to citizens.

Everyone would spend that savings in ways other than dealing with crime. Residents and businesses would have the means (saved money) along with the motive (occupy their own castle) to create safe and pretty places. By how much would American site value increase?

Apply the 10% above to GDP to get $2 trillion – formerly a cost and now a savings to spend. Since Americans usually spend 1/5th of their income on location, take 1/5th of $2tr, or $0.4tr – to get the increase in social surplus after cutting crime (in the streets, not in the suites).

II. Redevelopment

As residents feel safer, they fix up the formerly rundown neighborhoods. In doing so, they raise their property value. a tax liability). And businesses, too, spruce up Main Street, malls, and shopping centers. In a virtuous cycle, these improvements further cut crime. (I know, so why tax improvements?)

Attracted by the more hospitable environment, newcomers move in. More population creates more demand and more competition among buyers. They bid up the price or lease-rent for sites that structures sit on. Today, that means gentrification. But tomorrow, when residents receive a share of local land value, it means a fatter pie to divvy up.

In the bigger picture, people on the move bid up land value where they land – but deflate land value when they leave. So, for a region or nation, these local changes might not figure into the overall total. However, newcomers usually also pump up density. More consumers per acre increases business, increases income (means), increases prestige (motive), and thereby increases neighborhood site value.

Cities have been adding more people per acre, and ground rents just love them some crowds. Manhattan, with 70k residents per square mile, has a price tag of $700 billion for its 23 square miles of land. Even experts get it: "housing prices increase more strongly in cities with more severe supply constraints, as measured by higher population density ..."

To accommodate the newcomers, builders build. Where does the infill go? On vacant lots. On lots supporting abandoned and outdated buildings. The metro land needing such a makeover is at least a quarter of urban land – a lot of lots.

As builders increase supply, they don't knock down price, but actually raise it. For the half decade 2010-2015, the ratio was 7 to 1. When builders increased the housing stock 1%, they increased the housing (er, location) price 7%.

Another researcher said redeveloping urban acreage raises its value 25%. Wouldn't it be nice if we could get those two talking to each other? Who's ever right, let's forge ahead.

How much does total rent rise? To be conservative and make the math easy, apply 12% (between the 7% and 25% above). Apply it to the total we found for metro land value. It comes to about $0.6t. Add it to the subtotal we reached by cutting crime, $0.4t. Thus infill, plus density, on many metro acres, raises our near future increase to $1 trillion.

III. Transportation

Currently, to avoid emptying vaults of cash for a prime location, some choose to use the cheapest sites available, in areas desolate and sparsely settled. But do they really save? Over a century ago, a guy name Johann Heinric von Thunen noticed the cost of distance.

The farther "out" you go, the cheaper land gets, but the more expensive transportation gets. At the extreme, land is free (Death Valley). Versus: The farther "in" you go, the spendier land gets, but the cheaper transportation gets. At the extreme, transportation is free—in the heart of the city you can walk everywhere. You can plot it on a graph, drawing a diagonal line cutting the graph quadrant into two right triangles; the ratio is neatly 1 to 1.

Either way, you got to pay. But who does the math? Not us mathphobes (Ch 1). So, to accommodate those who cannot afford a heavy monthly rent or mortgage and choose to be nickeled-and-dimed while getting about in the boonies, government lays asphalt – and paves the way for sprawl.

Sprawl is fine for some. Politicians are eager to please the Growth Machine. Builders, pavers, realtors, lenders, and investors are the ones who deliver the fattest campaign contributions. Pay the piper, call the tune. And music to the ears of business is the ka-ching of coin spent on transportation – a $1.3 trillion annual expenditure, much of which pavers capture. Their cohorts in the Growth Machine benefit from having the way paved for shippers and potential customers.

The Growth Machine enriches a few a lot and taxes everyone somewhat. Perhaps you watched a new bridge being erected, or a new light-rail go in (a public investment), then checked out the new cost of housing or offices (a private investment). Higher, right? Thanks to the ease of access to those locations, making them more desirable. So, residents and businesses bid up the value of land. Again, a giving, just the opposite of a taking.

Yet, who connects the dots? So far, a minority. For the majority, it's easier to tolerate reality, pay taxes, and go with the flow – or get stuck in traffic. While the aim of paving new roads and widening old ones is greater mobility, they don't deliver like we want them to. They enable growth, which delivers traffic jams. It's a downward spiral that never ends.

Actually, not "never." Already, localities have started to cater more to riders, less to drivers, and add in the alternatives to driving. Sidewalks, bus lanes, bike paths, etc., all reduce traffic and raise mobility. Once residents receive shares of rent, more will choose to spend less time and money driving and more on being closer to the center of things with access to alternatives.

To serve an influx to downtown, where would developers put the new condos and shops? On land now devoted to cars: dealer lots, junk yards, gas stations, repair shops, insurance offices, patrol HQs, traffic courts, etc., and mainly overly wide streets and parking. The automobile's "tire-track" (not humanity's footprint) then shrinks drastically in compact cities.

To serve the growing ranks of urban dwellers, urban authorities do more to make their jurisdictions more livable, with bike lanes, pocket parks, pedestrian malls, trolleys, etc. In such cities, minus choking traffic, people don't feel crowded, even in high density. Instead, people feel pleased, and spend more to be there. How much more?

Walkways, bikeways, and open space are amenities people pay more for.

- In residential areas, improving a neighborhood's Walk Score (0 to 100) by only 15 points increased home-site price by an average of 12%. ("The Economic Value of Walkability: New Evidence" by Joe Cortright, *City Commentary*, 30 August 2016

- In walkable shopping areas, rents can be 27-54% higher than in non-walkable commercial areas. (Patrick Sisson at Curbed, 24 June 2019)

- Home buyers are willing to pay a premium of $9,000 to be within 1,000 feet of access to a bike trail. (Patrick Sisson at Curbed, 24 June 2019)

- Bike lanes in various cities raised nearby site values by 2% to 20%.

- Trails and greenways are amenities people pay more to be near. Seattle's Burke-Gilman Trail increased the value of homes near the trail by 6.5%. In Boulder, Colorado, the average value of a home adjacent to the greenbelt was 32% higher than a similar house 3,200 feet from the greenbelt.

- Both urban parks with open space bordering settled areas – and forested areas where trees grow – raise the value of nearby lots, the latter by 9.0%.

Setting aside land for nature and lanes for muscle power is not the only way to make metro regions livable. There's also ... Legalizing Uber. Self-driving cars. Express mass transit (whether bus or rail). Tunnels. Beyond transportation there's also ... Recycling of water, garbage, and trash. Non-polluting power sources. Plazas. Chiseled buildings at corners. Amphitheaters. Parks. Pocket parks. Daylighting streams. Wildlife corridors. This list is not exhaustive. However, my search failed to turn up any scholarly correlations of such improvements to land value.

Imagine all American metro regions adopting these amenities and more (as they've already started to do). People do spend more to experience life in a livable city. How much more could metro land command? To be conservative and simplify the math, let's use a middle value from above of 20%. One fifth of our current rent total for metro land ($5t) comes to $1t for a makeover of cities, leaving them walkable, bike-able, and livable, lets residents have fun, be healthy, and be kind to the environment. Add it to $1t for redevelopment and going crime-free, bringing our future bump-up to $2 trillion.

Yet there's cash left dangling. What happens to the $1 trillion that localities and residents now spend on sprawl annually? If citizens get feisty, their elected officials will have to either give it back or cut taxes. Either way, it leaves voters with more money to spend. The usual 1/5th to spend on location, $0.2 t, puts the future rise at $2.2 trillion.

IVa.

It's easy to misspend Other People's Money (OPM, pronounced "opium"). It's just human nature. So, what if you unsuccessfully spend the dough; OPM is a never-ending flow. In the public sector, revenue is always OPM. In the private sector, all the money that bankers and stockbrokers handle is OPM. Look how they redecorate their offices, flit about in company jets, when they should be increasing dividends. Conversely, when the money you spend is the money you worked for, you're far less likely to waste it.

Over time, the waste – and the interference into ordinary living that it funds (think airport body searches) – gets tiresome for many. Then the return on taxes is both too little and too irksome. So, voters want change. Some consider cutting down on government spending and public "services."

Enter formerly poor citizens now getting a dividend from regional land values. It's like an admission ticket to the middle class. Like most people enjoying greater income, recipients become owners and feel higher self-esteem. In the middle class, residents participate more in civic affairs. More residents vote, and even show up at city hall to argue about what to spend public dollars on and what to cut off from the public trough.

They could very well defund traditional programs; society would no longer need such services so much. Without the needy, we could cut charitable services. With crime becoming a rare aberration, we could shrink police budgets. By compacting cities, we could cut the underwriting of sprawl. Since fewer people would be joining up to get the GI benefits, we could trim the military. In general, we could diminish or eliminate the addictive subsidies and the counterproductive taxes that fund such programs.

How would streamlining the flow of public revenue impact land values?

IVB.

No matter what program of public spending you like, or which type of tax you don't like, all taxes and subsidies have four intrinsic, unavoidable flaws.

1. Taxes and subsidies cost. They do not just happen, they need to be staffed. You pay taxes to politicians who deliver revenue to bureaucrats who hire providers of various services who deliver their services to some. Those middle layers—IRS and state and local collectors, accountants, and enforcers—need to get paid. Each $1 they take as tax costs $0.67. Add on the hundreds of agencies and departments for food stamps, Indian affairs, Medicaid, schooling (apart from the schools themselves), energy (apart from actually delivering energy), NASA, the Pentagon, etc. The layers of bureaucracy are gigantic and expensive; some say 70% of the public budget stays in the building, only 30% reaches the supposed beneficiary. Even without waste like cheating military contractors, it costs money to take money to give money.

2. Taxes and subsidies distort price. Taxes make goods more expensive, which the sales tax makes obvious, but all taxes decrease your purchasing power. And subsidies make "bads" more affordable; note how the free money to agri-business makes high-fructose corn syrup artificially cheap and ubiquitous. That's the direct distortion. Indirectly, the things that politicians do not tax become relatively more affordable, such as lawyers. The things they do not subsidize become relatively less affordable, such as organic food. The whole economy becomes less efficient. Because government makes corn cheap, you eat too many cornflakes and not enough fresh fruit and spend too much on health care. Not being omniscient, politicians can never levy a tax or allocate a subsidy without conferring advantages and disadvantages. But, of course. Why else would anybody lobby?

3. Taxes and subsidies violate *quid pro quo*. No matter how much benefit you receive, or how much impost you pay, if the two equate, it is only a matter of luck and not likely to happen again in several lifetimes. Normally, taxes paid don't match dollar for dollar with subsidies received. If they did, there'd be no reason for lobbyists. The whole *raison d'etre* for lobbyists is to skirt taxes

and/or amplify subsidies by a return of 10, 100, or even 10,000, or – hold onto your hat – 76,000% over the amount invested in lobbying. Receiving such huge sums, those who win in the corridors of power ensure that others lose in the marketplace. Even when the losers could supply the consumer with better goods and services, businesses that can't lobby well get undercut, and you lose, too.

4. Finally, taxes and subsidies reinforce an old bugbear, *might makes right*. All of us, except insiders, have no choice but to pay taxes or go to jail – and to accept subsidies or miss out on stuff that's already been paid for. Being coerced, people use this fact to rationalize dishonesty: making inaccurate statements on forms, cheating customers in business, etc. Further, being coerced, everyone but insiders feels somewhat helpless. Some personality types drop out of civic life and make do with crumbs. Other types show excessive loyalty to the state.

Those applying the coercion – politicians, lobbyists, super wealthy, and aristocratic families – do all in their power to keep the gravy train rolling to the trough they slurp from, no matter how much suffering it causes others. They need to disable their conscience, which not only puts society at risk but also their own mental health. None of these responses to the inherent coercion of taxation and subsidization is healthy for individuals or social progress.

However unfair and inefficient taxes and subsidies are, having been around for millennia, we're used to them and largely incognizant of alternatives. Yet taxes are not a "necessary evil" (if that's not too strong for unfairness and inefficiency), and subsidies not necessary at all. Instead of taxes, governments could utilize the non-coercion-based fiscal tools of fees, leases, and dues, such as land dues. And instead of subsidies, lawmakers could simply pay you a dividend from social surplus directly, enabling you to hire the teacher or doctor or whomever you want.

A judge – a lawyer living off others paying taxes – called taxes the price we pay for civilization (more like for *domestication,* the libertarians among us would auto-correct). How high is that price? The four inherent deficiencies of taxes and subsidies drain away trillions of dollars. Plug the leaks (as with dues and dividends), redirect the savings to the populace, and they'd spend a goodly portion on bidding up the rent for locations. Let's see how much.

IVc.

How much are our elected officials and office holders wasting? Harvard says the federal government alone wastes $1 trillion. Then there are the states, counties, and cities. One pro-big-business, anti-big-government writer puts the total for all governments in America at $6 trillion – about 1/3 of GDP.

Imagine replacing subsidies to special interests with dividends to citizens. Government gets hugely downsized. Middle people, all the bureaucrats above, would need to find truly productive jobs or launch businesses. Either way, they'd quit dragging down the GDP and expand it instead.

When governments return the savings from no waste, or lower taxes, that $6tr above goes into the pockets of citizens. Those lucky ducks would spend the typical one-fifth on location, or $1.2tr. Human-friendly mobility, better development, and reduced crime had added $2.2tr, so now we're up to $3.4 trillion fresh rent in years to come.

Economists euphemistically dubbed waste due to lobbying as "rent seeking." More honestly it is "revenue winning." Whatever the name, it costs everyone else downstream. In any given year, persons who can afford to seek favor waste somewhere between $1 and $3.5 trillion in national output.

So, when government pulls the plug on them, they haven't the wherewithal to waste, and everyone else has that income to spend. (those trillions may or may not somewhat overlap these trillions, the cost of bureaucracy in the private sector: $3tr) People would spend an extra $0.2tr to $0.7tr on location. To round off, let's hang with the high side and use $0.6. Tack it onto the $3.4, bringing new rents to $4 trillion.

V.

Taxes and subsidies are especially hard on the environment. Politicians tax labor, making labor-intensive industries – mostly the "green" ones – less profitable than the capital-intensive "grey" ones – upon whom politicians confer loopholes. Businesses engaged in recycling, reforestation, organic farming, and solar energy all lose market share to competitors engaged in strip mining, clear-cutting, factory farming, and oil drilling. Additionally, politicians both subsidize fossil fuels and limit their liability, so providers of clean energy, not putting others at risk, lose a big competitive advantage.

With citizens serious about saving public dollars in order to swell their dividend, they'd target the handouts to polluters and depleters – and make those corporations pay their way. Not only would the environment heal but so would humans. They'd feel less stressed by toxins and by financial pressure.

People could save trillions on medical bills. Presently Americans spend about $3.6tr on doctors, prescriptions, hospitals, insurance, etc. Once we're well nearly full-time, we could save much of that expense, maybe $3tr. Because people use extra income to bid up land values (about 1/5th the savings), it's another $0.5tr. Now you're looking at extra land value reaching $4.5 trillion.

VI.

Since taxes and subsidies inhibit economies from growing, their absence would free economic grow. Turn from subsidies, gifts to insiders, to taxes – grabbing from outsiders. Eco-losses are not the only human activity (or depravity?) depressing land values. Among others are the taxes that shrink their base.

- The tax on income reduces income. The income tax discourages some from working harder, others from investing at all.

- The tax on sales decreases sales. The sales tax, by making goods more expensive, means some consumers cannot afford to buy quality products, shriveling prosperity.

- The tax on buildings lowers their value. The property tax, falling mainly on buildings, keeps some owners from making improvements.

Mason Gaffney and Richard Noyes compared those US states that rely more heavily on property taxes with those relying more heavily on income and sales taxes. The ones levying land, via the property tax, proved more prosperous than those burdening sales. It stands to reason.

While most taxes shrink their base (and should, in the case of a tax on pollution), not all do. Somewhat counter-intuitively, the property tax shift – un-taxing buildings while up-taxing land – actually grows its tax base. Not taxing improvements allows owners to build and improve without penalty. Meanwhile, falling on locations, it falls on what was already created before humanity arrived. Having to pay the tax does not motivate anyone to produce less land or hide it offshore. Rather, having to pay the levy spurs owners to build and improve. Meeting demand, that increases location value, the new tax base.

For the libertarians lurking among us, please note the tax aspect is not key. A fee or lease or land dues could work just as well. Now I feel better.

Nic Tideman, a former presidential advisor, with his grad student Florenz Plassman, calculated the result from replacing dumb taxes that shrink their base with one smart one falling on location (that expands its base). *Such a tax shift would increase GDP by nearly 30%.* That is, for the USA. In Uncle Sam's domain, taxes are relatively low. In countries with higher taxes, the forgone gains reach over 90% of their GDP. Anyway ... Current US GDP would go from a bit over $20+ trillion to about $26 trillion. Spending a fifth of that extra $6tr, or another $1.2tr as rent, pushes up the increase in future natural value to $5.7 trillion.

VII.

It's not easy to tell, but actually the cost of living keeps falling. We can't see that reality, due to the inflation of prices. Politicians and land ("home") buyers keep borrowing and bankers keep issuing more new notes than the economy produces in new goods and services. The excess cash gets used by lucky recipients to bid up the prices of their purchases – usually assets like stocks, bonds, REITs, and real estate – triggering a chain reaction.

Currently, lenders have leverage over borrowers, so they can charge interest. But could they charge so much, or at all, on a level playing field? Presently, most people must pay rent to live in someone else's house or apartment, very few of us get paid to house-sit somebody else's lodging. But what if capital were plentiful and savers needed to keep savings safe? We'd all enjoy the leverage of house-sitters. Then lending might not be profit making at all.

All the borrowing and going into debt has a solution.

- On the public side, curtail the discretionary spending of politicians. Just let citizens rely on rent dividends. Then politicians have no justification for over-borrowing.

- On the private side, as the land dues go up, the land prices must come down; the more land rent that the society gets, the less the seller and lender can get. So, mortgages shrink tremendously.

Without so much debt in the economy, bankers cannot inflate the money supply, so prices would stabilize.

Only briefly. The ongoing progress in technology would constantly batter costs, shrinking prices.

- Lower costs let people bring to market new ideas while lower prices let other people buy more goods and services, so GDP goes up.

- Liberated from financial worry, many people will pursue projects without expecting to get paid, such as counting monarch butterflies in Mexico to see how the species is doing. So GDP would go down.

As costs fall, as prices fall, as wages and profits fall, the one thing that would stay high relatively (even if not absolutely) is the value of location. In the near future, such spending could rise from the current $11t, more or less, by the $5.7 to $16.7t total. With GDP a bit over $20t, rent could be as much as 80% of GDP, the *Pareto optimum*. We'd be spending four times as much for locations and resources and on government-granted privileges like patents and copyrights than we'd spend on another's labor or capital. Sorry to hog so much of GDP. As Steve Martin said, *Excuuuuse me*!

As natural value piles up, it reveals the autonomous nature of wealth. Indeed, economies can't help but spew forth a surfeit of goods and services. All that's needed, really, is for…

- labor and capital to operate on the best locations (land),

- in an efficient public space, i.e., unfettered by excess regulation or counterproductive taxation (or subsidies to competitors), and

- in a fair private space, i.e., free from extortion of their earnings by nefarious pirates, legal or otherwise.

Then, presto! Bounty flows.

The Excess that Keeps on Exceeding

Those trillions spent for Earth would return to us as rent shares. While your expenses would keep falling, falling, falling, your "Citizen's Dividend" would swell, swell, swell, at least relatively, maybe absolutely. Earth's worth in America, soon to be $16t annually per capita of registered voters, comes to about $120k annually. Monthly, your share is $10,000, every month. Gee, that's like being rich. If you thought the work week could shrink and leisure expand with only $1k per month, imagine your lifestyle with $10k per month! We'd all be perfectly well off.

Wow. Oversize me. It's staggering. The size of this Citizen's Dividend is so huge, it's hard to process. Such gargantuan numbers make one feel awe. And feel doubt; it sounds too good to be true.

Think of earlier outrageous claims that turned out to be true. Harvests hundreds of times greater thanks to fertilizer. Hearing aids. Safe birthing after a mortality rate of 50%. Viagra. Dick Tracy wrist radio in everyone's hands. 50 mpg cars mass produced worldwide. Etc. Add to it the sheer bulk and dominance of rent.

It actually makes sense. In this case, the numbers don't lie. Just follow the steps we took to reach this grand total. It's not the product of ideology but of logic. Locations and privileges do command the overwhelming portion of our spending. Most of our income would no longer come from work or saving or investing; 80% of it would be our share of all these rents.

In actuality, most work does not directly put food on the table or a roof over one's head or create clothes, cars, or computers. Most work is not productive but conformist. Check out *The World's Wasted Wealth* by J.W. Smith to see how many jobs are performed to qualify for an income, not to create goods or services.

Our whole understanding of economies would change. The popular erroneous notion of labor vs. capital as the basic dynamic must fade away, to be replaced by the view of the struggle between rent-winners and rent-losers. Sharing natural bounty becomes the new normal, as normal as working at a job is now.

A good number for the worth of Earth in America could help people become aware of rent and understand how it comes about automatically. If humanity could get its ducks in a row and share rents, life on Earth becomes exquisite as it should be. That'd be reason to celebrate.

CHAPTER 42

CELEBRATE KNOWING RENT WITH A POWWOW

Howling success or not, having succeeded at all it's time to howl, Kimosabe.

POTLATCH PARTY TIME

For anyone with an inkling of interest in economic topics, having reached our goal calls for a toast. Now we know the worth of Earth in America (Ch 37). From official statistics, we squeezed a total for the bounty of the economy now, and how much it could be soon (Ch 41). Plus, putting our figure in a timeline paves the way to knowing the phase of the economy. Party time, eh?

We also saw how redirecting the flow of our spending for assets never created by labor and capital resolves major issues, both economic and ethical. Sharing the socially generated value of land could resolve competing claims over ownership. That'd give those afflicted by today's gentrification and yesterday's colonization reason to celebrate.

Conflicts between newcomers and old stayers are so ingrained as to seem intractable. Yet the same could have been (and was) said about institutionalized male dominance over females and a little longer ago about slavery. Today, around the world, both behaviors are almost universally despised.

If a community does share its land value, then, as new people move in, competing for housing, pushing up location value, they'd fatten each person's share, a residential dividend. People already settled there would have extra income to pay the newly raised land dues or land tax. Or, residents may take the money and run, to a neighborhood in the region where land dues (or taxes) are lower. At that location, their rent dividend would stretch much further.

Unaware of this economic solution, many residents argue instead for a political attempt (rent control, tax cap, etc) that usually turns out to be a non-solution. That approach is anti-market, while pro-market is the bias of many fellow Americans. One jurisdiction, however, eschewed command and instead tapped demand; Aspen, CO recovers a slice of local rent (the annual value of the region's natural assets) to help families afford desirable housing (Ch 40).

CATCH UP TO INDIANS

Longer-term residents who're anti-newcomer or anti-gentrification argue that tenure gives them rights and privileges that go with a title. Perhaps unknowingly, they provide a sound argument, "First come, forever crowned," for original inhabitants, too. Whoever reached a place first would be the rightful owners, with their descendants entitled to stay forever.

When one says "my land," one can refer to one's own land. Or to the land in their region. Using the latter meaning, those whose ancestors were displaced by colonizers want that land returned. Imagine descendants of Indians getting back Manhattan. Or Arabs getting back Palestine. Or Aborigines getting back Australia. Hard to imagine, without a real apocalypse.

Furthermore, for some individuals, getting a parcel of land would be less useful than receiving a monetary equivalent, a share of "rent." Hence many descendants of indigenous peoples, both in America and Australia, ask descendants of European settlers to pay a rent as reparations for historical and ongoing displacement. We've shown that there certainly is enough rental surplus for that purpose.

Critics raise several arguments against so-called reparations. What about mixed races? Partial payments? Whose math? And are natives going to pay reparations to even older nations whom they'd displaced? And are sons guilty of the sins of their fathers? Nobody alive now was alive then. Finally, the right to land is universal, so people have the right to join others anywhere on Earth.

In America, some native tribes have sued in court to have old treaties honored. Others demand back royalties and interest. To know how much compensation is owed, they need to know today's value of their lost land. During the pre-trial deposition, the attorneys would declare their competing estimates of the worth of that corner of the earth. While government does not keep a current total of all rent at its fingertips, a plaintiff with standing might be able to get government to dig down and 'fess up. During a trial, a federal judge might order the bureaucracy to measure natural America's worth.

Some Indians have actually collected. In 1996, Elouise Cobell, a Native American businesswoman from Montana, sued the federal government for underpaying royalties for more than 100 years. She won. In 2009, Congress and the president approved a $3.4 billion settlement for land that was held in trust by the government and never, until then, reimbursed in full.

Treaties aside, that's a drop in the bucket compared to what royalties would be for all native homeland. Try trillions. Think New York. Silicon Valley. Texas oil fields.

MANHATTAN FOR BEADS

Manhattan is one land purchase by early European settlers familiar to every schoolchild. But the facts surrounding the takeover may invalidate it.

Four centuries ago, Dutch settlers paid some Indians a pocketful of beads for Manhattan. However, the tribe who dealt with the Dutch, the Canarsee, actually lived in Brooklyn and probably were happy to accept hi-tech art for an island that belonged to others. Further, they did not sell the land. They couldn't. Hunter/gatherers conceived property differently. Law professor G. Edward White saw the Indians *"not relinquishing the island, but simply welcoming the Dutch as additional occupants."* Massasoit, leader of the Wampanoag when the Pilgrims arrived at Plymouth in 1620, said, *"What is this you call property? It cannot be the earth, for the land is our mother, nourishing all her children, beasts, birds, fish, and all men."*

The Dutch should've known better. As good Christians, they were theoretically prohibited from buying land. In their Bible, their God said, *"Thou shalt not own the land forever, for the land is Mine, you are My tenants."* (Leviticus 25:23) Wise Solomon claimed that, *"The profit of the earth belongs to all."* (Ecclesiastes 5:9). Further, the New Testament quotes Jesus saying, *"The meek shall inherit the earth."* (Matthew 5:5) On the meekness scale, those accepting trinkets surely rank higher than those proffering trinkets.

Since the Indians did not sell Manhattan but merely "leased" it, and ours is a litigious society, imagine descendants of the original inhabitants trying to collect so many trillions as their due. They'd need an authoritative source to cite, whether government or business or academic. Already, three researchers have estimated the selling price of Manhattan. At $1.4 trillion, the rate of return since the Dutch settlement in 1626, nigh four centuries ago, would be 6.4%. Since land price is a cumulative projection of land rent, the actual value might be $700 billion annually.

Whatever the tribes demand, when all residents of the region receive a rent dividend, aborigines would receive a quasi-compensation. It'd amount to a greater financial gain for most native descendants than for colonial descendants. Most Indians live in rural areas, most descendants of colonists live in cities. In the countryside, the cost of living is low; in cities, it's high. The dividend – the same size for everyone – would go much further in rural communities than in urban neighborhoods.

COUNTING CELEBRATES A NEW WORLDVIEW

If society shares Earth by sharing her worth, that can be an ideal way to resolve competing claims to the same location. The legitimate owners (on the basis of being first), receiving a share of regional values, might drop demands for returning lost values. Then people could enjoy and celebrate both prosperity and peace in their time.

The Indian cause, if including a rent grand total, could shift the paradigm. The distributions of rent to residents in Alaska, Aspen, and Singapore exist because people see those values as windfalls; Alaska's oil, Aspen's site value, and Singapore's budget surplus. What would it take for any region to see their worth as a windfall?

Count it. Show its immensity. Generate the contradiction of an embarrassment of riches. Most citizens don't know how huge the surplus is. Learning that fact, they'd be overawed, too. And feel that the current way surplus gets hogged is far out of bounds. And see that the size of rent is great enough that sharing; that compensation of others for displacing them, and being compensated for being displaced, is right and feasible.

Naturally, one identifies with and feels proud of their community and region. Its natural value – long ago seen as a common heritage – would be something to brag about; comparisons would be made. The swelling emotion could lead to sharing, the essence of *community*.

The root term in "community" is "mun," meaning share ("com" means with). It's the act of sharing that helps create one's identity with others. To share the commonwealth would not be to redistribute it but to *predistribute* it, before an elite or state has a chance to misspend it. Every monthly check would be reason for natives, urban dwellers, and just about everybody else, to celebrate.

When might we quit hoarding? Old paradigms would not be old if they were easy to replace. Yet they do change. Kuhn explains that new generations have not yet closed their minds to new ideas; some become early adopters. Today's young people – and not just those burdened by student debt – care about inequality and injustice in the rewards dished out by the economy.

Passionately caring souls do not constitute an active majority, but it only takes a critical mass. When the stars align, a movement reaches that threshold. Then, counting what's actually ours, becomes a *cause celebre.*

CODA

LET TIMES SQUARE CELEBRATE A NEW HIGH

The opposite of progress is Congress – 'cept for a gathering of geonomists.

WE CONCLUDE ...

Imagine the crowds in Time's Square celebrating not just a new year but a new number, a new high in the worth of Earth in America. It's the densest location in America and the most valuable; and the former leads to the latter.

Since 2009, Times Square, the heart of Manhattan, which is the heart of New York City, which is the heart of urban America (proof: the map gracing the cover of *The New Yorker*), has become more like any other world-class city, going car-free, setting up sidewalk cafes. Even with more shoppers (especially in the summer), the value of land there continued its recessionary slide for a couple years. The bigger they are, the harder they fall.

In actuality prices were still sky-high. But since 2011, values there have gone ballistic. Indeed, throughout much of the Big Apple, site values are in the stratosphere.

Because where the most people want to play – in city centers – that's where people pay the most to live and work. Densely populated New York, London, Tokyo, et al, have the priciest real estate on earth. Buyers are billionaires.

Their stratospheric values are all due to location, to being situated in a prosperous nation engaged in nonstop, vigorous trade. For these United States, we tallied all the metro, natural, spectral, etc., values we could lay our hands on. Our project did collect, collate, and package stats in intelligible form.

WHAT'S ACCOMPLISHED

An indicator for wasted land, for economic bounty, and for coming conditions is knowledge society can put to good use. We announced the tally to those who appreciate the effort to measure the value of all land plus privilege:

- readers here, obviously, and those who helped along the way,
- the foundation Schalkenbach pitched in,
- the politician California Rep. Mike Thompson speaks knowledgeably and implored Congress's research agency
- professional helpers, like cooperative sources as within the Fed itself,
- other agencies,
- the hundreds who got cited earlier. Beyond them,
- the agencies who should be doing this.

Beyond them, despite further concentrated media, all outlets—old print newspapers and magazines and 21st century e-media websites—using the most searched relevant words and visual memes for a press release, we disseminated summaries to communicators:

- business reporters,
- economic commentators,
- editors of print media and websites; TruthOut publishes us on occasion,

They're a start. Others will spread the word about a well grounded figure:

- activists for economic reform,
- visionary businesses like Zillow, and
- * investment consultants.

Still a minority but perhaps a critical mass.

Along the way, we discovered the most searched relevant words and gathered visual memes; essential ingredients for a TED Talk. Let it become the next intellectual *tour de force* following on the heels of Thomas Piketty's *Capital in the 21st Century*. Ours would be more like *Land Rents in the 21st Century*, but too few would grasp the special meanings of those technical terms. *Counting Bounty* feels better.

Broadcasting the news may cause a groundswell that reaches a critical mass who call for an official calculation of the worth of Earth in America. Now we wait, watching as readers and visitors connect the dots. One may feel like the first one at a school party to dance – awkward. But those who know what's fun, or are the affirming type, go from the sidelines to gung-ho. Soon almost everyone is dancing, even if it's an odd place to dance and all the participants are strangers.

PROGRESS

With more investigative resources, one could focus in more exactly. To continually refresh the statistic, one could DIY. Could we be the caretakers of the mission? Follow in the footsteps of NBER, Lincoln, and Zillow, which once calculated land value? Or, having blazed a trail, leave the responsibility for tracking all rents in the best possible hands? Those who'd keep the effort alive until academia and officialdom inevitably take over.

As Mahatma Gandhi did not say, "First they ignore you, then they laugh at you, then they fight you, then you win." Gandhi did write something similar albeit longer (Freedom's Battle, 1922). The quote is actually from a speech by labor leader Nicholas Klein to a textile union in 1914. The adage neatly expresses the stages of social reform.

1st, those curious about the worth of Earth are being profoundly ignored. They can only hope soon will come ...

2nd, the ridicule—they'd be one major step closer to their goal.

3rd, the elite and minions would oppose research fiercely; their last gasp might to indebt everyone, as by siphoning ever more frequent bailouts from public treasuries, most recently for the covid lockdown—again good news. Right around the corner ...

4th, the quest to know the tally for America's own assets becomes popular and after that, finally official.

Phew. Then society could finally get this genomic show on the road.

Paradigms shift, but not of their own accord. First, somebody has a new view. As Kuhn noted, somebody must come from outside the relevant field (as where we stand). The outsider's new analysis makes sense. It gets resisted. But it still makes sense, so it spreads and eventually wins. Why wait?

Questions? Comments? Ideas? Please get in touch. Let's have fun! Instead of dropping, maybe the Times Square ball could just keep rising.

###

Index